Projects for Calculus
the language of change

SECOND EDITION

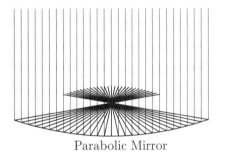 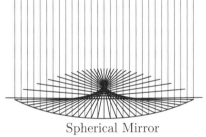

Parabolic Mirror and Spherical Mirror

Projects for Calculus
the language of change

SECOND EDITION

Keith D. Stroyan
Mathematics Department
University of Iowa
Iowa City, Iowa

Typeset with $\mathcal{A}_{\mathcal{M}}\mathcal{S}$-TeX

Academic Press
San Diego   London   Boston   New York   Sydney   Tokyo   Toronto

This book is printed on acid-free paper. ∞

Copyright © 1999, 1993 by Academic Press

All rights reserved.
No part of this publication may be reproduced or transmitted in any form or by any means, electronic or mechanical, including photocopy, recording, or any information storage and retrieval system, without permission in writing from the publisher.

Mathematica is a registered trademark of Wolfram Research, Inc.
Maple is a registered trademark of Waterloo Maple, Inc.

ACADEMIC PRESS
a division of Harcourt Brace & Company
525 B Street, Suite 1900, San Diego, CA 92101-4495, USA
http://www.apnet.com

ACADEMIC PRESS
24–28 Oval Road, London NW1 7DX, UK
http://www.hbuk.co.uk/ap/

Library of Congress Cataloging-in-Publication Data

Stroyan, K. D.
   Projects: Calculus: the language of change/Keith D. Stroyan —2nd ed.
      p. cm.
   Rev. ed. of: Calculus using Mathematica. c 1993.
   Designed to accompany: Calculus: the language of change/Keith D. Stroyan 2nd ed. c1998.
   ISBN 0-12-673031-8
   1. Calculus—Computer-assisted instruction. 2. Calculus—Data processsing. 3. Mathematica (Computer file) 4. Maple (Computer file) I. Stroyan, K. D. Calculus using Mathematica. II. Stroyan, K. D. Calculus. 2nd ed. IIII. Title.
QA303.5.C65S77   1998 Suppl.
515'.078'553—dc21   98-19464
                                                   CIP

Printed in the United States of America
98 99 00 01 02 IP 9 8 7 6 5 4 3 2 1

CONTENTS

1 INTRODUCTION TO THE SCIENTIFIC PROJECTS

0.1	What Good Is It?	2
0.2	How Much Work Are They?	2
0.3	How to Write a Project	2
0.4	Help on the World Wide Web	3
	PROJECT 1. Linear Approximation of CO_2 Data	5

2 EPIDEMIOLOGICAL APPLICATIONS

1.1	Review of the S-I-R Model	10
1.2	Basic Assumptions	11
1.3	Derivation of the Equations of Change	11
	PROJECT 2. The 1968-69 New York Hong Kong Flu Epidemic	13
	PROJECT 3. Vaccination for Herd Immunity	17
3.1	Herd Immunity	18
3.2	The Contact Number Data	18
3.3	Project Issues	19
3.4	Vaccine Failures	19
	PROJECT 4. S-I-S Diseases and the Endemic Limit	21
4.1	Basic Assumptions	21
4.2	The Continuous S-I-S Variables	22

v

Contents

4.3	Parameters for the SIS Model	22
4.4	The Importance of the Contact Ratio	23
4.5	Conjectures	24
4.6	Conclusions	25
	PROJECT 5. Max-Min in S-I-R Epidemics	27

3   THE ROLE OF RULES FOR DERIVATIVES

	PROJECT 6. The Expanding Economy	33
	PROJECT 7. The Expanding House	35
7.1	Volume Expansion Explained by Calculus	38

4   APPLICATIONS OF THE INCREMENT APPROXIMATION
$$f[t + \delta t] - f[t] = f'[t]\,\delta t + \varepsilon \cdot \delta t$$

	PROJECT 8. A Derivation of Hubble's law	43
	PROJECT 9. Functional Linearity	47
	PROJECT 10. Functional Identities	51
10.1	Additive Functions	54

5   DIFFERENTIAL EQUATIONS FROM INCREMENT GEOMETRY

	PROJECT 11. The Tractrix	57
	PROJECT 12. The Isochrone	59
12.1	Conservation of Energy	61
	PROJECT 13. The Catenary	63
13.1	The Catenary Hypotheses	63
13.2	Parameters	63
13.3	Variables	64
13.4	The Equation for Tension	64
13.5	Optimizing Length and Strength	68

Contents

6   LOG AND EXPONENTIAL FUNCTIONS

Project 14.	The Canary Resurrected	73
Project 15.	Drug Concentration and "Biexponential" Functions	75
15.1	Primary Variables of the Model	76
15.2	Parameters of the Model	76
15.3	The Formulas for Concentration	77
15.4	Comparison with Mythical Data	80
15.5	Comparison with Real Data	80
Project 16.	Measurement of Kidney Function by Drug Concentration	81
16.1	Variables and Parameters	81
16.2	Overview of the Project	81
16.3	Drug Data	82
Project 17.	Numerical Derivatives of Exponentials	87
Project 18.	Repeated Exponents	91
Project 19.	Solve $dx = r[t]\, x[t]\, dt + f[t]$	

7   THEORY OF DERIVATIVES

Project 20.	The Mean Value Math Police	97
20.1	The Mean Value Theorem for Regular Derivatives	97
20.2	The Theorem of Bolzano	99
20.3	The Mean Value Theorem for Pointwise Derivatives	100
20.4	Overall Speed IS an Average	101
Project 21.	Inverse Functions and Their Derivatives	103
21.1	Graphical Representation of the Inverse	104
21.2	The Derivative of the Inverse	105
21.3	Nonelementary Inversion	111
Project 22.	Taylor's formula	113
22.1	The Increment Equation and Increasing	114
22.2	Taylor's formula and Bending	115
22.3	Symmetric Differences and Taylor's formula	116
22.4	Direct Computation of Second Derivatives	118
22.5	Direct Interpretation of Higher Order Derivatives	118

Contents

8 APPLICATIONS TO PHYSICS

PROJECT 23. The Falling Ladder (or Dad's Disaster)	123
23.1 Air Resistance on Dad (Optional)	125
PROJECT 24. Falling with Air Resistance: Data and a Linear Model	127
24.1 Terminal Velocity	129
24.2 Comparison with the Symbolic Solution	130
PROJECT 25. Bungee Diving	131
25.1 Forces Acting on the Jumper before the Cord Is Stretched	131
25.2 Forces Acting on the Jumper after He Falls L Feet	133
25.3 Modeling the Jump	135
PROJECT 26. Planck's Radiation Law	137
26.1 The Derivation of Planck's law of Radiation	138
26.2 Wavelength Form and First Plots	138
26.3 Maximum Intensity in Terms of a Parameter	141
PROJECT 27. Fermat's principle Implies Snell's law	145
27.1 Reflection of a Curved Mirror	147
27.2 Computation of Reflection Angles	149

9 APPLICATIONS IN ECONOMICS

Project 28. Monopoly Pricing	157
28.1 Going into Business	157
28.2 Going into Politics	160
PROJECT 29. Discrete Dynamics of Price Adjustment	161
29.1 The Story	161
29.2 The Basic Linear Model	164
29.3 Taxation in the Linear Economy (Optional)	164
29.4 A Nonlinear Economy (Optional)	165
29.5 Taxation in the Nonlinear Economy	166
PROJECT 30. Continuous Production and Exchange	169
30.1 Why Trade?	169

Contents

10 ADVANCED MAX-MIN PROBLEMS

PROJECT 31. Geometric Optimization Projects	175
31.1 Distance between Lines	175
31.2 Distance between Curves	177
31.3 An Implicit-Parametric Approach	178
31.4 Distance from a Curve to a Surface	179
PROJECT 32. Least Squares Fit and Max-Min	181
32.1 Introductory Example	181
32.2 The Critical Point	182
32.3 The General Critical Equations	183
PROJECT 33. Local Max-Min and Stability of Equilibria	185
33.1 Steepest Ascent	185
33.2 The Second Derivative Test in Two Variables	186

11 APPLICATIONS OF LINEAR DIFFERENTIAL EQUATIONS

PROJECT 34. Lanchester's Combat Models	193
34.1 The Principle of Concentration	195
34.2 The Square Law	195
34.3 Guerrilla Combat	196
34.4 Operational Losses (Optional)	197
PROJECT 35. Drug Dynamics and Pharmacokinetics	201
35.1 Derivation of the Equations of Change	203
35.2 Where Do We Go from Here?	210
35.3 Periodic Intravenous Injections (Optional Project Conclusion)	210
35.4 Steady Intravenous Flow	213
35.5 Intramuscular Injection - a Third Compartment	214

Contents

12 FORCED LINEAR EQUATIONS

PROJECT 36. Forced Vibration - Nonautonomous Equations	217
36.1 Solution of the Autonomous Linear Equation	218
36.2 Transients - Limiting Behavior	218
36.3 Superposition for the Spring System	218
36.4 Equations Forced by Gravity	220
36.5 Equations Forced by Sinusoids	221
36.6 Nonhomogeneous IVPs	223
PROJECT 37. Resonance - Maximal Response to Forcing	225
37.1 Some Useful Trig	225
37.2 Resonance in Forced Linear Oscillators	226
37.3 An Electrical Circuit Experiment	227
37.4 Nonlinear Damping	227
PROJECT 38. A Notch Filter - Minimal Response to Forcing	229
38.1 The Laws of Kirchhoff, Ohm, and Coulomb	229
38.2 Steady-State Solution	231
38.3 A Check on a[t] 2	233
38.4 Where's the Min?	233

13 APPLICATIONS IN ECOLOGY

PROJECT 39. Logistic Growth with Hunting	237
39.1 Basic Fertility	237
39.2 Logistic Growth	238
39.3 Voodoo Discovers the Mice	239
PROJECT 40. Predator-Prey Interactions	241
40.1 Bunny Island	241
40.2 Rabbit Island	245
PROJECT 41. Competition and Cooperation between Species	247
41.1 Biological Niches	247
41.2 Cooperation between Species	251
PROJECT 42. Sustained Harvest of Sei Whales	253

Contents

42.1	Carrying Capacity, Environmental and Mathematical	253
42.2	The Actual Carrying Capacity	255

14 DERIVATIONS WITH VECTORS

	PROJECT 43. Wheels Rolling on Wheels	261
43.1	Epicycloids	262
43.2	Cycloids	264
43.3	Hypocycloids	265
	PROJECT 44. The Perfecto Skier	267
44.1	The Mountain's Contribution	268
44.2	Gravity and the Mountain	270
44.3	The Pendulum as Constrained Motion	271
44.4	The Explicit Surface Case	271
	PROJECT 45. Low-Level Bombing	273
45.1	Significance of Vector Air Resistance	277
	PROJECT 46. The Pendulum	279
46.1	Derivation of the Pendulum Equation	280
46.2	Numerical Solutions of the Pendulum Equation	282
46.3	Linear Approximation to the Pendulum Equation	284
46.4	Friction in the Pendulum (Optional)	287
46.5	The Spring Pendulum (Optional)	287
	PROJECT 47. Using Jupiter as a Slingshot	291
47.1	Setting up the Problem: Scaling and Units	292
47.2	Newton's Law of Gravity	293
47.3	Newton's $F = ma$ Law	294
47.4	Numerical Flights out of the Solar System	294

15 CHEMICAL REACTIONS

	PROJECT 48. Stability of a Tank Reaction	297
48.1	Mass Balance	297
48.2	Arrhenius' Law	298
48.3	Heat Balance	298
48.4	Stability of Equilibria	300
48.5	Forced Cooling (Optional)	302
	PROJECT 49. Beer, Coke, & Vitamin C	303
49.1	Enzyme-mediated Reactions	303
49.2	Molar Concentration and Reaction Rates	303

49.3	The Briggs-Haldane Dynamics Approximation	305
49.4	The Michaelis-Menten Dynamics Approximation	306
49.5	Blood Ethanol	307
49.6	Blood CO_2	308

16 MORE MATHEMATICAL PROJECTS

PROJECT 50.	Rearrangement of Conditionally Convergent Series	313
PROJECT 51.	Computation of Fourier Series	317
PROJECT 52.	The Big Bite of the Subtraction Bug	323

17 ADDITIONAL PROJECT REFERENCES

1001.1	Lowering the Water Table	326
1001.2	The Light Speed Lighthouse	326
1001.3	Horizontal, Vertical, and Slant Asymptotes	326

Chapter 1

Introduction to the Scientific Projects

0.1. What Good Is It?

Most students are not interested in mathematics for just for mathematics' sake. They often don't see the importance of studying mathematics, even if their teachers have told them, "When you get to ____ you will need to know this (esoteric) mathematical trick." This sounds like, "When you grow up, you will be glad you factored your polynomials." (Something like eating spinach?) As far as we are concerned, you have grown up now. It is time for us to 'put up or shut up' as mathematics teachers. These projects show you why and how calculus is important in a number of deep examples. You cannot answer the questions raised in these projects without calculus and you can answer them with calculus.

The projects explore questions from serious ones like, "Why did we eradicate polio, but not measles?" to fun ones like, "Will the bungee diver rip his leg off or smash his head on the rocks?" There is a wide variety and we hope you will find several that interest you. If you want to know why you should study calculus read through the list of projects, then work one that is on a topic you find interesting.

The projects are an important part of our calculus course. Calculus is a beautifully coherent subject in its own right, but even if you are interested in studying mathematics for its own sake, you should be aware of the fact that calculus has important things to say about science and the world around us. Working a few of these projects over the course of the year will not only show you the power of using calculus to express "how things change," it will also help you understand the mathematics. (The Math Background book on the CD answers all your theory questions.)

0.2. How Much Work Are They?

The projects are not easy. They require a lot of work and time to think. You don't just plug in a formula from the book, manipulate for 3 minutes, and go home, but you will find the effort worthwhile. No pain, no gain. Without exception, students understand the point of these projects once they have worked them. A few find it 'too much work,' but almost all are justifiably proud of the effort that goes into a finished project.

At the University of Iowa, students (with majors from music to mechanical engineering) work three large projects in the first year. (Students at other schools have also worked many of these projects.) They choose topics of interest and spend up to 3 weeks on and off working out their complete solution with a partner. In the large projects, teams submit a first draft for criticism and partial credit after a week, then clarify and refine their work in a final report of about 10 pages. During the week before the first submission, we do not have lectures; student teams work in the lab.

0.3. How to Write a Project

The goal of the projects is for you and a partner to write a "technical report" or "term paper" on the subject. You should not expect to "find THE answer," write it, circle it, and hand it in to an instructor who already knows what it is. Rather, your final goal is a report. Imagine sending it to your boss at your new job. Your boss is no dummy, and she can fire you, but she has been out of school a long time and doesn't remember the details of her calculus course (or the subject of the project). You need to write a clear explanation of the problem, your basic steps in solving it, and your conclusions.

The projects have "exercises" aimed at helping you get started and posing the right questions. The exercises are not the whole story. In this edition I called them "hints" because some students treated them like the drill problems in high school math. They felt that if they just worked the exercises, with nose to the paper, they would be done. The hints are only intended to help you discover the "big picture." The hints become gradually more vague and sometimes build up to more open-ended "explorations." Some of these projects don't even have single answers, but require you to make choices.

Read the project introduction and work the exercise hints, but remember that the projects require you to develop an overall understanding of the subject. Organizing your work on the hints and writing a clear explanation of your solutions will help you develop this understanding.

Finally, write the "Introduction" to your report last. Do NOT begin your project by rewriting my introduction. Start by working on the hints. Do NOT write an introduction like, "In our project we are going to ...." But rather something like, "Using calculus, we found a simple mathematical criterion needed to prevent the spread of a disease. This report derives the condition and gives examples. In the case of polio, our condition is easily met, 20% of the population can be susceptible, whereas, for measles ..." Use your introduction as an advertisement to interest your reader in the rest of the report or, in the case of your busy corporate boss, to give her the basic conclusions clearly and concisely.

0.4. Help on the World Wide Web

Our website
$$\text{http://www.apnet.com/pecs/books/stroyan/}$$
contains Mathematica and Maple programs for each project to help you solve the hint exercises in the projects and to help you write your report. Check it out. The web is also a rich source of information about the projects.

PROJECT 1

Linear Approximation of CO_2 Data

This project uses high school mathematics (the equation of a line) and the computer to give you an idea of the the scope of the problems we call "Projects." There is also a calculus moral to the story. The main approximation of calculus is to fit a linear function to small changes of a nonlinear function. Large-scale linear approximation makes sense in some problems but works in a different way. There is a pitfall in using a "local" approximation to make a long term prediction.

Global warming is a major environmental concern of our time. Carbon dioxide (CO_2) traps heat better than other gases in the atmosphere, so more CO_2 makes the atmosphere like a "greenhouse." Increased industrial production of CO_2 and decreased conversion of CO_2 to oxygen by plants are therefore thought to be causes of a general warming trend around the world. This is the greenhouse effect.

The computer program "co2 program" contains monthly average readings of CO_2 similar to those taken at Mauna Loa Observatory, Hawaii. Figure 1.1 shows a graph of the program's data from 1958 through 1967. (We couldn't obtain the raw data, so approximated it. Our graph looks similar to the published data.) The CO_2 level oscillates throughout the year, but there is a general increasing trend that appears to be fairly linear. It is natural to fit a linear function to this general increasing trend.

FIGURE 1.1: CO_2 Data

One way to make a linear approximation is simply to lay a transparent ruler on the graph and visually balance the wiggles above and below the edge of the ruler. That is a perfectly valid mathematical start, but the computer also has a "Fit" function and it produces the "approximation"

$$f = 0.6783x - 1013.$$

The graph of the data and this fit are shown in Figure 1.2.

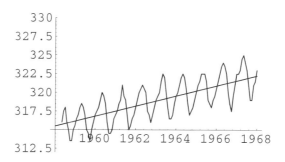

FIGURE 1.2: Computer Fit of CO_2 Data

The linear upward trend looks about like the yearly average of the data. The yearly averages are computed and fit to another approximating function

$$g = 0.7133x - 1081.$$

in the computer program **co2 program**. This is shown in Figure 1.3

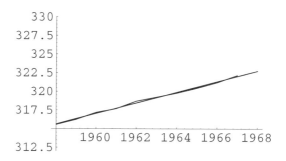

FIGURE 1.3: Yearly Averages of CO_2

Notice that the first fit to all the data is not the same as the fit to the annual averages of the data, but it is close. Both fits and sets of data are shown in Figure 1.4.

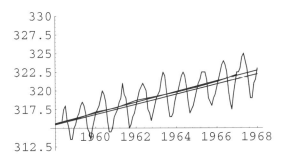

FIGURE 1.4: Data with Both Fits

HINT 1.5.
1) Use the "eyeball" method of putting a ruler on the graph of the data where it balances the wiggles and then write the formula

$$h = ax + b$$

for your approximation, $a = ?$, $b = ?$. You will notice that the "slope-intercept" formula $h = ax + b$ is not the most convenient. If your ruler is at the points $(1958, 315)$ and $(1968, 320)$ then the change in h is 5 (ppm) for a change in x of 10 (years). How would you compute a? What obvious point does the line

$$\frac{(h - h_0)}{(x - x_0)} = \frac{(\Delta h)}{(\Delta x)}$$

pass through? We need the slope-intercept form for comparison with computer, but the point-slope formula for a line is most convenient in this part of the exercise. What is the point-slope form of the equation of a line?

2) How does your approximation compare with the two computer approximations f and g above?

3) Use all three approximating functions f, g, and h to predict the 1983 CO_2 average. How much increase in CO_2 over the 1958 level is your prediction?

When we did the preceding exercise, our calculation of the increase in CO_2 level over the 1958 level was 17 ppm, for a level of 333 ppm. The measured average turns out to be 343 for an increase of 27 ppm. That is an error of $\frac{10}{27}$. The long-term data and prediction are shown in Figure 1.6.

The worrisome thing is that the average rate of increase from 1968 to 1983 is more than the linear trend of the 1958 to 1968 decade. What is your estimate of the present level of CO_2?

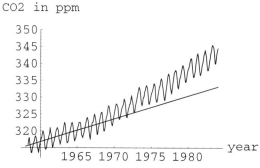

FIGURE 1.6: Linear Prediction of CO_2

You can read more in an article, Under the Sun - Is Our World Warming, in *National Geographic*, October 1990 (vol.178, nr.40). You can also use the computer to fit nonlinear functions to the data. For example, you might try a quadratic polynomial or exponential fit and use them to make predictions. Unfortunately, we do not know a scientific basis to choose a nonlinear function for our fit, so such predictions are only a curiosity.

Chapter 2

Epidemiological Applications

This chapter contains four projects. The first matches theory with empirical data, the second uses the proven model to make an important prediction, the third explores a change in basic assumptions and its consequences, while the fourth finds the "peak" of an epidemic. You can work the first three projects after you have studied Chapter 2 of the main text, the fourth after Chapter 11, or all later when you study Chapter 21. These are serious real applications of calculus, but they are not very hard in a technical sense (elaborate rules and formulas are not needed – thanks to the computer.)

The New York City Hong Kong Flu Epidemic

This project compares the S-I-R model with actual data of the Hong Kong flu epidemic in 1968-69. The data we have to work with are "observed excess pneumonia-influenza deaths." In this case our 'removed' class includes people who have died. This is a little gruesome, but it is difficult to find data on actual epidemics unless they are extreme in some way.

This project has programming help in the program **FluDataHelp** on our website, so you won't have too much trouble manipulating the data.

Vaccination Strategies for Herd Immunity

This project uses the mathematical model of an S-I-R disease to find a prediction of how many people in a population must be vaccinated in order to prevent the spread of an epidemic. In this project you use data from around the world to make predictions on successful vaccination strategies. Dreaded diseases like polio and smallpox have virtually been eliminated in the last two generations, yet measles is a persistent pest right here on campus. The mathematical model can shed important light on the differences between these diseases. The mathematics is actually rather easy once you understand the basic concept of decreasing infectives. Why did we eradicate polio, but not measles? Calculus gives a clear simple answer.

Endemic S-I-S Diseases

Some infectious diseases such as strep throat, meningitis, or gonorrhea, do not confer immunity. In these diseases, there is no removed class, only susceptibles and infectives. This project has you make a mathematical model for S-I-S diseases.

An S-I-S disease has the potential to become endemic, that is, approach a nonzero limit in the fractions of susceptible and infected people. The disease never goes away. The first problem in this project is to find mathematical conditions that say when there is an endemic limit,

$$\lim_{t\to\infty} s[t] = s_\infty \neq 0 \quad \text{and} \quad \lim_{t\to\infty} i[t] = i_\infty \neq 0$$

You will be able to do simulations by modifying the **SIRSolver** program or getting the help program from our website.

The "Peak" of an S-I-R Epidemic

There are various maxima associated with an epidemic; the largest portion of the population sick at one time, the fastest increase in the portion of sick people, and the fastest increase in new cases. Even though we cannot find explicit classical formulas for $s[t]$ and $i[t]$, we can still use calculus to find these "peaks."

1.1. Review of the S-I-R Model

Before working one of the the specific epidemic projects, you should review the basic assumptions that we made in building with our S-I-R model in Chapter 2 of the text. Write a derivation of the model in your own words as a first section of your project.

1.2. Basic Assumptions

We make the following assumptions about the disease in formulating our model:

(1) **S-I-R:** Individuals all fit into one of the following categories:
: Susceptible: those who can catch the disease
: Infectious: those who can spread the disease
: Removed: those who are immune and cannot spread the disease
(2) **The population is large,** but fixed in size and confined to a well-defined region.
(3) **The population is well mixed**; ideally, everyone comes in contact with the same fraction of people in each category every day.

1.3. Derivation of the Equations of Change

In order to describe the way individuals move between the susceptible, infected and removed compartments, we used two parameters, b and c, and the product $a = b \cdot c$ in our S-I-R model. The values are specific to a particular S-I-R disease.

PARAMETERS FOR THE S-I-R MODEL

$c =$ the number of contacts adequate to transmit the disease per infective

$b =$ the daily recovery rate, 1/period of infection

$a = b \cdot c$ the daily rate of contacts per infective

The value of b is observed clinically, since $b = 1/$the number of days an individual is infectious.

The parameter c is called the "contact number" because it gives the average number of close contacts each infective has over the course of the disease. The value of c is measured indirectly through the invariant equation given in Chapter 2 of the main text. Describe this measurement briefly in your project if you use measured values of c. Observed values of c are given in the herd immunity project below.

We measure the contact number $c = \frac{a}{b}$ and then compute the daily contact rate $a = c \cdot b$.

THE CONTINUOUS S-I-R VARIABLES

In order to develop a model that applies to different populations, we used fractional variables.

$t =$ time measured in days continuously from $t = 0$ at the start of the epidemic

$s =$ the fraction of the population that is susceptible $= \dfrac{\text{number susceptible}}{n}$

$i =$ the fraction of the population that is infected $= \dfrac{\text{number infected}}{n}$

$r =$ the fraction of the population that is immune $= \dfrac{\text{number removed}}{n}$

where n is the (fixed) size of the total population. Because of our assumptions,

$$r = 1 - s - i$$

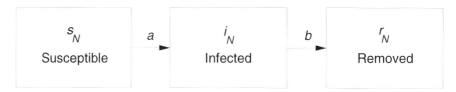

FIGURE 1.7: S-I-R Disease Compartments

THE S-I-R DIFFERENTIAL EQUATIONS

$$\frac{ds}{dt} = -a\,s\,i$$
$$\frac{di}{dt} = a\,s\,i - b\,i$$

These equations describe the rates at which people are getting sick and recovering. Each project in this chapter should begin with a derivation in your own words that explains how the terms $a\,s\,i$ and $b\,i$ describe these rates of change. See Chapter 2 for help. The S-I-S project will derive slightly different equations.

PROJECT 2

The 1968-69 New York Hong Kong Flu Epidemic

In order to verify our model of an S-I-R disease, we need to know the proportions of a population that are susceptible, immune, and removed as functions of time in a real epidemic. It is often very hard to determine these proportions because a large number of cases of a given disease go unreported (consider how many times you've had the flu, and out of those times, how many times you actually saw a doctor and had a chance to be "counted"). For this reason, certain diseases lend themselves to study because, in one way or another, they make their presence known to the medical community. For example, polio was severe enough to require medical attention, ensuring that there would be records of most cases. The Hong Kong flu epidemic data of 1968-69 were derived from a different source.

The data for the Hong Kong flu are from the medical reports of coroners in New York City. During the period of the epidemic, the disease can be traced as a function of the excess of individuals who died due to influenza-related pneumonia in the New York City area. The term "excess" in the preceding sentence refers to the fact that you expect to get a certain number of deaths. However, during the epidemic, there was a large increase in deaths which could be roughly attributed to the Hong Kong flu (due to influenza-related pneumonia). The data (from the U. S. Centers for Disease Control) follow and are also given in the computer program **FluDataHelp** on our website (so you don't have to retype them).

Once we have our data, the question becomes, how to compare the data to our model? There are two factors that we must consider. First, not everyone who got the Hong Kong flu died as a result. And second, those who did die probably did not do so right away, but rather had the disease for a period of time before expiring. On the basis of these two facts, then, we can hypothesize that the number of excess deaths is proportional to the number of infected individuals from a couple of weeks before.

A comparison with the model requires that we satisfy the assumptions of our model and set up the initial conditions. The population of New York in 1968 was around 7,900,000. Because the Hong Kong flu was a new strain of influenza at the time that the data were taken, all individuals were susceptible. Due to its great virulence, the disease doesn't require much help getting started – say 10 people who have the virus return from a visit to Hong

Kong, raising the population to 7,900,010.

Week	Excess Deaths
1	14
2	31
3	50
4	66
5	156
6	190
7	156
8	108
9	68
10	77
11	33
12	65
13	24

HINT 2.1. *Assumptions of the Model*
Although New York City is obviously not a static population, for the sake of simplicity we will represent it as such – as a result, the only way to get out of the Susceptible group is to become an Infective, after which you must enter the Removed category, which includes both deaths and recoveries. Begin your project write-up with a discussion of the basic assumptions that go into the derivation of the model:

(S-I-R DEs)
$$\frac{ds}{dt} = -a\,s\,i$$
$$\frac{di}{dt} = a\,s\,i - b\,i$$

where b is the inverse of the period of infection; a is the contacts per infective individual; and s, i, and r are fractions of the population in the various categories of infection. $r = 1 - s - i$. What are the values of the constants a, b, and c?

How well does New York City satisfy the assumptions, especially the "homogeneous mixing" assumption? Where does this enter the differential equations?

You can do computations with this model using either of the computer programs **SIR-solver** or **FluDataHelp** (from our website).

After scaling and allowing for time for the epidemic to start, we see the following superposition of model prediction and measured excess deaths shown in Figure 2.2.

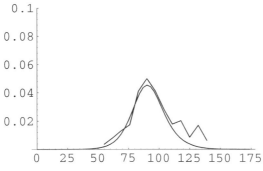
FIGURE 2.2: Theory and Data

Your job is to make a similar match and justify that your match verifies that the S-I-R differential equations do indeed make reasonable predictions about flu epidemics.

HINT 2.3. *Start the Model*
Open the computer program **FluDataHelp** and run the computation of the model with the starting values $s[0] = 7900000/7900010$, $i[0] = 10/7900010$ and the known parameters for flu, $b = 1/3$, $c = 1.4$, and $a = b \cdot c$. Calculate for 90 days at first, then decide whether the epidemic takes longer or shorter than 90 days. When does the number of infectives reach its maximum?

HINT 2.4. *Look at the Data on the Computer*
The data above are contained in the program **FluDataHelp**. They need to be converted to days and started at an appropriate time. The 10 original cases took some time to start a widespread epidemic, and only after that did we observe excess deaths. Change weeks to days and assume the death observations start after a number of days. How many days would you need to shift the data days so that the peak deaths occur at the same time as the maximum of infectives from the model? Graph the revised data.

HINT 2.5. *The Verification*
Scale and slide the data to make a reasonable comparison between the model and the data. Then JUSTIFY your scaling, writing the justification in your own words.

PROJECT 3

Vaccination for Herd Immunity

This project uses the mathematical model of an S-I-R disease to find a prediction of how many people in a population must be vaccinated in order to prevent an epidemic of that disease. Dreaded diseases like polio and smallpox have virtually been eliminated in the last two generations, yet measles is a persistent pest right here on campus. Why have we eradicated polio, but not measles? The mathematical model can shed important light on the differences between these diseases.

HINT 3.1. *Begin your project write-up with a description in your own words of how we derive the S-I-R differential equations. (See Chapter 2 of the text.)*

HINT 3.2. *Continue your project write-up by answering the question: How are a, b and c determined for a real disease? (See Chapter 2 of the text.) The values of c for various epidemics are given below, but you should explain how they might be determined for a new disease. Once these are measured for one epidemic, the values can be used to make predictions about another.*

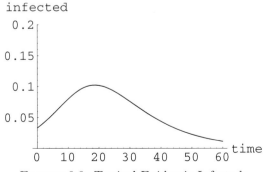

FIGURE 3.3: Typical Epidemic Infecteds

3.1. Herd Immunity

Up to this point, all of our disease models have dealt with a disease that spread through the population. However, there are cases in which the population can attain what is called "herd immunity." In such cases a high enough percentage of the population is immune to ensure that if the disease is introduced to the population, it will simply die out and not spread to form an epidemic. In order for the disease to die out, the number of infectives must be declining. The basic question we need to answer is: When will the infectives (i in our mathematical model) decrease?

HINT 3.4. *What mathematical condition (on some derivative) says that the number of infectives is declining? Use the model to express this condition strictly in terms of the susceptible fraction s and the contact number c. (At first, you will find a formula using s, i, a, and b, but you can simplify it to only use s and c.) If your condition fails, it means that the disease increases in the number of infected people. If it holds, i for the disease declines.*

	Ave. measured c	Ave. observed 1/b
Measles	16.3	8
Whooping Cough	16.5	
Chickenpox	11.3	
Scarlet Fever	8.5	
Mumps	8.1	
Rubella	7.3	11
Smallpox	5.2	
Polio	4.6	

FIGURE 3.5: Table of Contact Numbers

3.2. The Contact Number Data

Now, consider the case in which the fraction of infectives is zero. If this is the case, then we can solve for s in terms of r and substitute this into the equation we derived above. This gives us a relationship between the contact number and the fraction of recovered individuals within the population. The next exercise is the first big part of your project. Recall that the first big question is: Why did we eradicate polio, but not measles?

HINT 3.6. *Figure 3.5 is a table of diseases with their corresponding contact numbers. Determine the minimum fraction of recovereds that must be present in the population in order for herd immunity to occur. Use the condition for declining i from the previous exercise.*

3.3. Project Issues

Do the resulting fractions of immune individuals needed to acquire herd immunity comply with what you might expect? How can you rationalize your data? Why do we still have measles and rubella, but not polio and smallpox? Explain the relationship between the contact number and the fraction of immune individuals required for herd immunity in terms of the infectiousness of the disease - that is, use words, not mathematical symbols and formulas.

Of the diseases in the table, for which one does it appear hardest for a population to attain herd immunity?

What complications might exist in the real world when it comes to attaining the required fraction (hint: think about immunization and patient isolation techniques and how they may vary from one region of a country to another, and between countries)? In a free society, how much force can we use to require immunization?

In order to demonstrate just how problematic these complicating factors can be, consider the case of smallpox, which has the lowest contact number in our table. Smallpox vaccinations began in 1958, but the disease persisted until 1977. How could you explain this? More important, what implications does this have for the likelihood of a population attaining herd immunity against a disease with a larger contact number?

3.4. Vaccine Failures

Let's consider another complicating factor in reaching the required fraction of immune individuals. Vaccines are not always effective – a small fraction of individuals who are vaccinated will not become immune. About 15 to 20 years ago – a notable period in history – there were problems with measles vaccinations: the vaccines had a low "efficacy." The measure of a vaccine's effectiveness is its vaccine efficacy, VE. This is the fraction of vaccinations that result in immunity for the inoculated person. Using this new parameter and your previous equation, determine the fraction of individuals who must be vaccinated (assuming your population consists entirely of susceptibles prior to immunization) to achieve herd immunity for the above diseases.

Now let's consider two specific cases, rubella and measles, which both have a vaccine efficacy of .95. Using the contact numbers given previously, compute the fraction of individuals who must be vaccinated in order to incur herd immunity. What effect does the vaccine efficacy have on the likelihood of eradicating these diseases from the population? What are some techniques that could be employed to increase the population's chances of attaining herd immunity? (You may know that many universities require undergraduates to be revaccinated.)

The United States, rubella vaccination policy is an example of an attempt to vaccinate a high enough percentage of the population so that the disease will be held in check and, theoretically, herd immunity could be attained. In order to carry out this policy, laws have been enacted that require children to have a rubella immunization prior to entering school – it is now estimated that 98 percent of the children entering school have been vaccinated. As a result of this policy, the incidence of rubella in America has been steadily declining. Can herd immunity be attained with this fraction of children being vaccinated? Upon what other factor(s) does the herd immunity depend? (hint: the immunizations began in 1969). The typical vaccine in this case is the MMR – measles, mumps, rubella booster. From this fact, we can assume that about 98 percent of the children entering school have been immunized against measles, as well. Is this fraction high enough to attain herd immunity?

There are social consequences of these health policies. Some groups in our free pluralistic society do not believe in immunization. Some groups do not want to create records of their existence. How much social 'leeway' does each vaccination strategy have?

PROJECT 4

S-I-S Diseases and the Endemic Limit

In Chapter 2 of the text we built a model of S-I-R infectious diseases. That model corresponds to diseases in which there are three classes of people: susceptibles, infectives, and removeds. However, some infectious diseases do not confer immunity, such as strep throat, meningitis, or gonorrhea. In these diseases, there is no removed class; there are only susceptible and infective classes. Once you recover and are no longer infectious, you become susceptible again.

We would like you to make a mathematical model for S-I-S diseases and study it. As with any model, it has limitations. It will not give us exact answers, but it will help us to see the general trends and make some important predictions. We use a potential outbreak of strep throat on campus as a test case for our model.

4.1. Basic Assumptions

We will make the following basic assumptions:

(1) **SIS:** Individuals all fit into one of the following two categories:
   : Susceptible: those who can catch the disease
   : Infected: those who can spread the disease
   *Note: There are no removeds who are immune and cannot spread the disease.
(2) **The population is large** but fixed in size and confined to a well-defined region. We imagine the population to be in a large public university during the semester when relatively little outside travel takes place.
(3) **The population is well mixed**; ideally, everyone comes in contact with the same fraction of people in both categories every day. Again, imagine the multitude of contacts a student makes daily at a large university. (This assumption does not work well for gonorrhea – see below.)

Gonorrhea is an S-I-S disease, but it is not well modeled by simple single compartments for susceptibles and infectives. The added concept of a "core" group is needed to build accurate models. See the book *Gonorrhea Transmission Dynamics and Control* by H. W. Hethcote and J. A. Yorke, Springer Verlag Lecture Notes in Biomathematics, vol.56, 1984, if you wish to study that disease.

In order to build a model that makes predictions for all large well-mixed populations, we make our main variables the continuous fractional variables similar to the ones in Chapter 2 of the core text.

4.2. The Continuous S-I-S Variables

$$t = \text{time measured in days continuously from t=0 at the start of the epidemic}$$
$$s = \text{the fraction of the population that is susceptible} = \frac{\text{number susceptible}}{n}$$
$$i = \text{the fraction of the population that is infected} = \frac{\text{number infected}}{n}$$

where n is the (fixed) size of the total population. Note that $s + i = 1$.

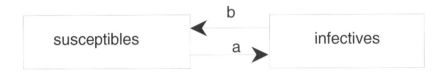

FIGURE 4.1: SIS Disease Compartments

4.3. Parameters for the SIS Model

$a = $ the number of contacts adequate to transmit the disease per infective per day

$b = $ the daily recovery rate, $= 1/\text{period of infection}$

$c = \dfrac{a}{b} = $ the contact number

As in Chapter 2, the parameter a determines the spread of the disease in the model. A serious challenge in this project is to decide how we might measure a or the contact number $c = \frac{a}{b}$. (This is not so hard once you figure out the limiting behavior.)

The average number of susceptibles infected by one infective is $a \cdot s$ per day. This is "adequate contacts times fraction of contacts susceptible." Therefore, if the entire infective class has size I, we get new infections at the rate $a\,s\,I$ per day, and since we work with the fraction of infected, this makes the rate of increase in i from new infections $a\,s\,i = a\,s\,I/n$.

The parameter that controls how quickly people recover and become susceptible again is b. You can also think of $1/b$ as the average number of days an infected person remains infective. For strep throat this is 10 days, so $b = 0.1$. As b is the recovery rate, the number of people leaving the infected compartment each day is $b\,I$ and the fractional rate is $b\,i$.

Let's look at the changes affecting the infected compartment. Each day the percentage of the population leaving the infected compartment is $b\,i$ and the percentage of the population that becomes newly infected is $a\,s\,i$. The differential equation for the infecteds is thus

$$\frac{di}{dt} = a\,s\,i - b\,i$$

Notice that this is the same as it was for the S-I-R model and represents the arrows in and out of the infected compartment in Figure 4.1.

You must now develop the differential equation for the susceptibles.

HINT 4.2. *The Rest of the Model*
Give the differential equation for the rate of change of the susceptible fraction. (Hint: Think of where the people leaving the infected compartment are going and where the ones entering the infecteds are coming from.)

You could compute i without a differential equation, if you knew s. Since everyone is either susceptible or infectious, $s + i = 1$ and $i = 1 - s$.

HINT 4.3. *Express the differential equation for s without using i?*

The last hint means that you can either use the system of differential equations for both s and i or use a single differential equation for s and solve for $i = 1 - s$ algebraically.

HINT 4.4. *The Computer and Your S-I-S Model*
Use one of the equivalent mathematical models from the previous exercises to modify the main computation of the computer program **SIRsolver** *to compute the course of an S-I-S epidemic. Run several test cases, say, using $b = 0.1$ and $a = 0.01$ and then $b = 0.1$ and $a = 1$.*

Use your program for initial experiments and later to verify conjectures.

4.4. The Importance of the Contact Ratio

The contact number, $c = a/b$, is the average number of adequate contacts an infective person has during the whole infective period. If the contact number is large, that intuitively means that the disease spreads more easily than a disease that has a lower contact number. The exercises that follow deal with seeing how the contact number affects the behavior of the disease in the model. This will lead you to make a mathematical conjecture about the importance of c in determining whether the disease persists in the population.

Imagine the following situation. Strep throat breaks out on campus. You catch it but are determined to go to class, thereby spreading it to some of your classmates. Your classmates are likewise conscientious and in turn pass it on. You recover but sit next to a third-generation infected person and catch it again.... In this way it might be possible for a certain fraction of the population to always be sick.

HINT 4.5. *The Endemic Limit – Experiments*
Use your the computer program (built from **SIRsolver** *in the last exercise) to try to formulate some conjectures about the importance of the contact number c.*

It seems reasonable that if a disease has a large contact number, that everyone becomes infected. Experiment with "large" (but fixed) contact numbers. Does

$$\lim_{t \to \infty} i[t] = 1$$

Why not? We suggest that you try $c = 2, 3, 4$. What does this mean in terms of spread of the disease?

Also experiment with "small" contact numbers. Try $c = \frac{1}{2}, \frac{1}{3}, \frac{1}{4}$. Be sure to compute a long time period such as a whole semester. What are biologically reasonable meanings of "big contact number" and "small contact number"?

If $\lim_{t \to \infty} i[t] = i_\infty$, what does the limit of $s[t] \to s_\infty$ have to be in your model?

You should have done enough experiments with the computer and your S-I-S model to see that for some values of c, the S-I-S disease dies out, whereas for other values of c, the disease tends to a limiting situation where it is always present. This does NOT mean that the same people are always sick. Why? This does mean that the disease is "endemic" in the population.

HINT 4.6. *Conjecture*
PART 1: *After you try various simulations, make a conjecture about*

$$\lim_{t \to \infty} s[t] = ?, \text{ when } c < 1$$

What do the numerical simulations do in the cases you ran when $c < 1$?

4.5. Conjectures

After you try various simulations, make a conjecture about

$$\lim_{t \to \infty} s[t] = a \text{ formula in terms of } c$$

in case $c > 1$. This formula should be easy to guess if you have done simulations with values such as $c = 2, 3, 4$ and compute your observed limit in terms of c.

Once you have made clear conjectures, you should use the equations for change to show that things change in the direction of your conjecture. For example, in case you conjecture that the disease dies out, $s[t]$ tends toward 1, so any initial condition with $s_0 < 1$ should eventually result in $s[t]$ increasing toward 1.

HINT 4.7. *Proof*
In the limiting case where $i[t] \to i_\infty \neq 0$ and $s[t] \to s_\infty < 1$, what happens to the rate of change of i with respect to t when $i = i_\infty$? What happens to the rate of change of s with respect to t when $s = s_\infty$?

 (1) Express your answer in terms of derivatives.
 (2) Express your answer in terms of the formula for derivatives given by the model.
 (3) Solve the second expression for $s = ?$

Suppose that you happened to start the solution of your model differential equation with $s[0] = s_\infty$, the limiting value. What would happen? What does this mean in terms of your differential equations?

Prove your experimental conjecture mathematically and say how the contact number determines whether or not an S-I-S disease will be endemic.

What happens if s is greater than the third expression?

What happens if s is smaller than the expression?

What happens if s cannot be greater than the last expression? (Recall that s is a fraction, so that $s \leq 1$.)

Summarize your results as a completely stated theorem, something like, "An S-I-S disease approaches an endemic limit, (s_∞, i_∞), provided the contact number ... " Beginning at any fraction of susceptibles $s[0]$ between zero and one, the limiting behavior is for $s[t] \to \cdots$ and $i[t] \to \cdots$.'

4.6. Conclusions

If you knew that an S-I-S disease was endemic and you knew the period of infectiousness, $1/b$, how could you measure the parameters a and c?

Now that you have explored the effects of the contact ratio on the long-term behavior of the disease, what could be done to help control the disease? What policies could be implemented by Student Health, for example, to help keep students infected with strep throat from becoming endemic? Say how the health policy decision would affect the parameters of your model and how much the effect would need to be to prevent endemic presence of the disease.

EXPLORATION 4.8. *Extensions of the S-I-S Model (optional)*

If a person with strep throat receives antibiotic treatment, after a day he is no longer infectious. On the other hand, he is not susceptible either. How would antibiotic treatment at an average of 3 days after infection change your model? How does the endemic limit theorem work in your new model?

PROJECT 5

Max-Min in S-I-R Epidemics

We never found explicit formulas for the functions $s[t]$ and $i[t]$ of the epidemic model of Chapter 2, and this project shows how you can find max-min information anyway.

The epidemic model of Chapter 2 is given by the system of differential equations

(S-I-R DEs)
$$\frac{ds}{dt} = -a\,s\,i$$
$$\frac{di}{dt} = a\,s\,i - b\,i$$
$$r = 1 - s - i$$

and the values of the fractions of susceptible, infective, and removed individuals at the start of the epidemic. For example, we studied the case $s[0] = \frac{2}{3}$ and $i[0] = \frac{1}{30}$. Typically, the initial $i[0]$ is small and $s[0]$ is large but not 1 unless no one has ever had the disease.

HINT 5.1. *Begin your project write-up with a description in your own words of how we derive the S-I-R differential equations. (See Chapter 2 of the text.)*

HINT 5.2. *Prove that the function $s[t]$ is a decreasing function on $[0, \infty)$. What calculus criterion is needed?*

There are two 'peaks' of interest in an epidemic. 1) When are the most people sick? 2) When is the disease spreading fastest? The second peak could mean either that the rate of contracting new cases is largest or that the growth in the infectious population is largest. These are different.

Before you begin, review the mathematical ideas needed to find the maximum of a function defined on the interval $0 \leq t < \infty$. You need to apply the theory to three functions, $i[t]$, $f[t] = a\,s[t]\,i[t]$, and $g[t] = \frac{di}{dt}[t]$.

HINT 5.3. *Suppose $h[t]$ is a differentiable function. We find*

$$\text{Max}[\,h[t]\ :\ 0 \leq t < \infty\,]$$

by . . . .

For example, the function $i[t]$ has a "typical" slope table, "up"-"over"-"down."

$$\frac{di}{dt} = a\,s\,i - b\,i = a\,i\left(s - \frac{b}{a}\right) = a\,i\left(s - \frac{1}{c}\right)$$

so

$$\frac{di}{dt} = 0 \Leftrightarrow 0 = a\,i\left(s - \frac{1}{c}\right) \Leftrightarrow i = 0 \text{ or } s = \frac{1}{c}$$

Moreover,

$$\frac{di}{dt} = a\,i\left(s - \frac{1}{c}\right) > 0 \Leftrightarrow i > 0 \text{ and } s > \frac{1}{c}$$

and

$$\frac{di}{dt} = a\,i\left(s - \frac{1}{c}\right) < 0 \Leftrightarrow i > 0 \text{ and } s < \frac{1}{c}$$

Note that $s[t]$ is decreasing, so if $s[0] > 1/c$ and later $s[t_p] = 1/c$, afterward we must have $s[t] < 1/c$ for $t > t_p$.

HINT 5.4. *The First Peak of an S-I-R Epidemic*
When is the epidemic expanding in the sense that the number of sick people is increasing? If you are thinking in terms of absences from class, when does the epidemic "peak"? Write your condition using a derivative and then express your answer in terms of s and the contact number $c = \frac{a}{b}$. Show that the variable increases before your condition and decreases afterward.

Consider various cases of c and initial conditions in s and i including "extreme" cases such as c very big and very small or $s[0]$ very big or very small. (Note: $c = 15$ for measles and $c = 4.6$ for polio. Nearly everyone is susceptible if $s[0] = 0.9$.) When $s[0] < 1/c$, when does the $i[t]$ "peak"?

HINT 5.5. *The Second Peak of an S-I-R Epidemic*
When is the disease spreading fastest in terms of the growth of new cases? What does $a\,s\,i$ measure? If we want to maximize the function $f[t] = a\,s[t]\,i[t]$ for $t \geq 0$, we need to find zeros of its derivative

$$\begin{aligned}\frac{d(s \cdot i)}{dt} &= \frac{ds}{dt} \cdot i + s \cdot \frac{di}{dt} \\ &= (-a\,s\,i)\,i + s\,(a\,s\,i - b\,i) \\ &= (a\,s\,i)\left(s - i - \frac{b}{a}\right)\end{aligned}$$

Show that this is zero when $s - i = 1/c$. When is $f[t]$ increasing? When is $f[t]$ decreasing? What is the slope table of its graph?

The previous peak does not take the daily recovery rate into account.

HINT 5.6. *The Third Peak of an S-I-R Epidemic*
When is the disease spreading fastest in terms of the growth of infectives? In other words, maximize the function $g[t] = \frac{di}{dt} = a s i - b i$. Show that the derivative of $g[t]$ is:

$$\begin{aligned}
\frac{dg}{dt} &= \frac{d(a s i - b i)}{dt} \\
&= a \frac{d((s - \frac{1}{c}) i)}{dt} \\
&= a^2 i \left(\left(s - \frac{1}{c}\right)^2 - s i \right)
\end{aligned}$$

This equation is hard to analyze, but we can use the invariant from Chapter 2,

$$s + i - \frac{1}{c} \text{Log}[s] = k, \qquad \text{a constant}$$

with the computer to find the crossing points of the curves

$$s + i - \frac{1}{c} \text{Log}[s] = k \quad \& \quad s^2 - s i - \frac{2}{c} s + \frac{1}{c^2} = 0$$

or crossing points of the curves

$$s + i - \frac{1}{c} \text{Log}[s] = k \quad \& \quad s - i = \frac{1}{c}$$

Discuss various cases of the value of c and the initial values of s and i.

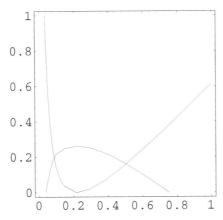

FIGURE 5.7: Intersection of the Critical Equation and Invariant

We can also modify the **SIRsolver** program to plot the expressions for $f'[t]$ and $g'[t]$ in various special cases.

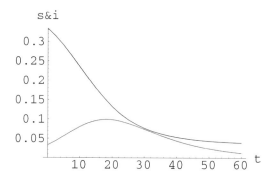

FIGURE 5.8: Susceptible and Infectious Fractions

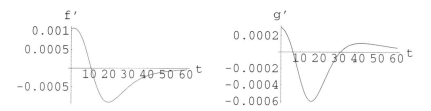

FIGURE 5.9: Derivatives of $f[t]$ and $g[t]$ from SIRsolver values of $s[t]$ and $i[t]$

These graphs should help you make slope tables for $f[t]$ and $g[t]$ at least in these special cases. (Look at the signs of $f'[t]$ and $g'[t]$ on the graphs.) Of course, you could also graph $f[t]$ and $g[t]$ themsleves.

EXPLORATION 5.10. *Use the program **SIRsolver** or **SIRmaxHelp** from our website to examine the three "peaks" of various epidemics. When do they occur? How do the disease parameters b and c affect the peaks and times of occurrence? How do the initial values of $s[0]$ affect the peaks and times?*

Chapter 3

The Role of Rules for Derivatives

The projects in this chapter are completely independent. The expanding house requires no calculus until the very end, where it is related to the chain rule, while the expanding economy is a practical illustration of the product rule. Both of these are short projects.

PROJECT 6

The Expanding Economy

This project is a small project or big exercise that illustrates the idea of the product rule for differentiation.

The population of a certain country is $15,000,000$ people and increasing at the rate of $10,000$ people per year. In this country the per capita expenditure for energy is $\$1,000$ per year and is growing at the rate of $\$8$ per year. ("Per capita" means each person spends $\$1,000$ each year.) This project asks you to explore the connection between the country's growth in total yearly energy expenditure and the symbolic product rule.

First do some arithmetic to get a feel for the problem. What will the population of the country be next year? How many dollars will each person spend on energy next year? What will the total energy cost be next year? How much has the total energy cost changed from the first to the second year? What are these same quantities the following year?

Now choose appropriate variables and carefully express the conditions of the problem in terms of your variables and their changes. For example, let P equal the population at time t (in years) and let E equal the per capita expenditure at time t. The total energy expenditure then is $T = P \times E$. These are all functions of time, $P = P[t]$, $E = E[t]$, and $T = T[t]$.

HINT 6.1. *For a change in t of one year, $\Delta t = 1$, denote the change in population by ΔP. Denote the change in per capita energy expenditure for a single year by ΔE. What is the change in total energy expenditure, $\Delta T = T[t+1] - T[t]$, written as a formula in terms of P, E, and the differences ΔP and ΔE? Illustrate your formulas with numerical examples.*

The symbolic question is more difficult, so here are some hints: The change in total energy expenditure for the people $P[t]$ during the next year is $P[t] \times \Delta E$, the difference in their total expenditure in year t, $P[t] \times E[t]$, and the next year, $P[t] \times (E[t] + \Delta E)$,

$$P[t] \times (E[t] + \Delta E) - P[t] \times E[t] = P[t] \cdot \Delta E = P \cdot \Delta E$$

However, this is not all of the new energy expenditure, because there are also new people, $P[t+1] - P[t] = \Delta P$. These people spend last year's amount on energy, $\Delta P \times E$, and they also spend the increase in per capita energy, $\Delta P \times \Delta E$. What is the formula for the one year change in total energy expenditure in terms of P, ΔP, E, and ΔE,

$$\Delta T = \text{old peoples' new energy}$$
$$+ \text{ new peoples' old energy}$$
$$+ \text{ new peoples' new energy}$$

Check your formula against the arithmetic you did at the beginning of the project and show the substitution for each term explicitly.

HINT 6.2. *What is the size of the new peoples' new energy compared with the sum of the other two terms in the expression for ΔT?*

Now we want you to modify your formula for the energy change ΔT if the time change $\Delta t < 1$, say $\Delta t = \frac{1}{12}$, one month. The number of new people in one month is the annual rate times $\frac{1}{12}$ and in Δt years is the annual rate times Δt. Let ΔP now denote the change in population during the Δt time period. Since the rate of change of population is assumed constant, we could express this as

$$\Delta P = (\text{change in } P \text{ during } \Delta t) = \frac{\Delta P}{\Delta t} \cdot \Delta t = \text{the annual rate times the time}$$

Similarly, the amount of new per capita energy expenditure in a time difference of Δt is

$$\Delta E = (\text{change in } E \text{ during } \Delta t) = \frac{\Delta E}{\Delta t} \cdot \Delta t = \text{the annual rate times the time}$$

HINT 6.3.

(1) *Show that the amount of new total energy expenditure is*

$$\Delta T = P \cdot \Delta E + \Delta P \cdot E + \Delta P \cdot \Delta E$$
$$= P \cdot \frac{\Delta E}{\Delta t} \cdot \Delta t + \frac{\Delta P}{\Delta t} \cdot E \cdot \Delta t + \frac{\Delta P}{\Delta t} \cdot \frac{\Delta E}{\Delta t} \cdot \Delta t^2$$

Explain each of these summands in terms of changes in people and energy.

(2) *Use the previous expression to give the rate of change of total energy during Δt,*

$$\frac{\Delta T}{\Delta t} = P \cdot \frac{\Delta E}{\Delta t} + \frac{\Delta P}{\Delta t} \cdot E + \frac{\Delta P}{\Delta t} \cdot \frac{\Delta E}{\Delta t} \cdot \Delta t$$

Explain each of these summands as rates of change.

(3) *Why is the third summand in the last expression the smallest?*

(4) *Write the general Product Rule expression for the instantaneous rate,*

$$\frac{dT}{dt} = \frac{d(P[t] \cdot E[t])}{dt} = ? \cdot \frac{dE}{dt} + ? \cdot E$$

Why is this the correct instantaneous formula for the rate of change of total energy expenditure? (That is, what happened to the third summand?)

PROJECT 7

The Expanding House

We want to investigate the question: How much does the volume of a typical house increase over the course of a normal day's warming? This problem illustrates a number of basic ideas which are generalized in symbolic differentiation formulas, and the numerical answer may surprise you, besides. Do you think it increases by a thimble, a bucket, or a bathtub?

Most solids expand when they are heated in familiar temperature ranges. (Ice near freezing is a notable exception.) Scientific tables list "coefficients of expansion" in units of 1/(units of temperature). Here are a few in units of $1/°C$:

aluminum	25.0×10^{-6}
copper	16.6×10^{-6}
gold	14.2×10^{-6}
iron	12.0×10^{-6}
wood	4.0×10^{-6} approx.

The fact that different solids expand at different rates makes some interesting engineering problems in structures built from different materials.

The first question is: What does the 'coefficient of expansion' mean? The units are actually (units of length expansion) per (units of length to start) per (units of temperature). The length units cancel. If the coefficient of expansion is c, the object is initially l units long, and the temperature increases ΔT °C, then the change in length

$$\Delta l = c \times l \times \Delta T$$

so a piece of wood 40 feet long at freezing (0° C) warmed to 61° F (= 16° C) increases its length by

$$\Delta l = 4 \times 10^{-6} \times 40 \times 16 = 2.56 \times 10^{-3} \text{ ft} = 0.0307 \text{ in}$$

This isn't very much, but how much would a house expand in volume?

Let's say the interior of the house is a rectangular solid 40 feet long, 30 feet wide, and 8 feet high. The changes in width and height are

$$\Delta w = 4 \times 10^{-6} \times 30 \times 16 = 1.92 \times 10^{-3} \text{ ft} = 0.0230 \text{ in}$$

$$\Delta h = 4 \times 10^{-6} \times 8 \times 16 = 5.12 \times 10^{-4} \text{ ft} = 0.00614 \text{ in}$$

The original volume of the house is

$$V = l\,w\,h = 40 \times 30 \times 8 = 9600 \text{ cu ft}$$

and the heated volume is

$$V = l\,w\,h = 40.00256 \times 30.00192 \times 8.000512 = 9601.84 \text{ cu ft}$$

The important quantity for our original question is the change or difference in volume

$$\Delta V = \text{heated volume - original volume} = 1.84 \text{ cu ft}$$

so the rate of change is

$$\frac{\Delta V}{\Delta T} = \frac{1.84}{16} = .115 \text{ cu ft/degree}$$

HINT 7.1. *Which is closest to 1.84 cubic feet in volume, a thimble, a bucket or a bathtub?*

HINT 7.2. *A copper pipe runs down the 40 foot length of your house before warming. How much does it stick out the end of the house afterward? How much force would be required to compress it back to the edge of the house? (You will need to look up a strain constant for copper for the second part.)*

HINT 7.3. *How much does a gold palace the same size as your house expand its volume?*

These numbers are interesting in that small linear expansions still produce a fairly big volume expansion, but all the arithmetic is not very revealing in terms of showing us the rates of expansion. A symbolic formulation can do this for us. We begin by specifying our variables:

ΔT = the change in temperature in degrees Celsius
l = the length in feet as a function of temperature
w = the width in feet as a function of temperature
h = the height in feet as a function of temperature
V = the volume in cubic feet as a function of temperature

where we begin with $l_0 = 40$, $w_0 = 30$, $h_0 = 8$, and $V_0 = 9600$ at the initial temperature, and c is a parameter for the coefficient of expansion. We know that $V = l\,w\,h$, but l, w, and h each depend on the temperature change ΔT. The coefficient of expansion formula $\Delta l = c \times l \times \Delta T$ gives us the formulas

$$l = l_0 + \Delta l = l_0(1 + c\Delta T)$$
$$w = w_0 + \Delta w = w_0(1 + c\Delta T)$$
$$h = h_0 + \Delta h = h_0(1 + c\Delta T)$$

so that the volume becomes

$$\begin{aligned} V &= l\,w\,h \\ &= l_0(1+c\Delta T) \times w_0(1+c\Delta T) \times h_0(1+c\Delta T) \\ &= l_0\,w_0\,h_0\,(1+c\Delta T)^3 \quad = \quad V_0(1+c\Delta T)^3 \\ &= 40 \times 30 \times 8 \times (1+c\Delta T)^3 \quad = \quad 9600(1+c\Delta T)^3 \end{aligned}$$

We can think of this expression as the composition of two simpler functions

$$x = x(\Delta T) = (1+c\Delta T)$$

the multiplier of a linear quantity needed to get its heated length and

$$V = V(x) = V_0\,x^3 = 9600\,x^3$$

The quantity x determines linear expansion, which in turn determines volume.

HINT 7.4. *Give direct computations like those from Chapter 5 of the core text to show that a small change in volume as x changes to $x + \Delta x$ is*

$$\Delta V \approx V_0 3x^2 \Delta x = 9600 \times 3x^2\,\Delta x$$

so that this change of Δx produces the ratio or rate of change

$$\frac{\Delta V}{\Delta x} \approx \frac{dV}{dx} = 3\,V_0\,x^2 = 28800\,x^2$$

Hint: Compute $\Delta V = V(x+\Delta x) - V(x) = \cdots$ symbolically.

Now use the definition of the variable x.

HINT 7.5. *Show that the change in x is $\Delta x = c\Delta T$ as temperature changes from 0 to ΔT, so*

$$\frac{\Delta V}{\Delta T} = \frac{\Delta V}{\Delta x}\frac{\Delta x}{\Delta T} \approx 3\,V_0\,x^2\,c$$

Finally, consider the size of x^2 compared with the other terms.

HINT 7.6. *Show that*

$$\frac{\Delta V}{\Delta T} \approx 3V_0 c$$

and

$$\Delta V \approx 2.88 \times 10^4 \times 4 \times 10^{-6} \times \Delta T = .115\Delta T$$

Hint: Calculate x^2 numerically.

We wanted $\frac{\Delta V}{\Delta T}$ and approximated it in two steps, $\frac{\Delta V}{\Delta x} \times \frac{\Delta x}{\Delta T}$. The chain rule says in general that $\frac{dV}{dT} = \frac{dV}{dx}\frac{dx}{dT}$ when V is "chained" together from two other functions.

HINT 7.7. *Explain why the final expression*

$$\frac{\Delta V}{\Delta T} \approx 3 V_0 c$$

shows how the formula produces a large volume expansion from a small linear expansion, while at the same time the length of the house is only changed a tiny amount using the formula

$$\frac{\Delta l}{\Delta T} = l_0 \cdot c$$

Now give a "practical" application of your formula.

HINT 7.8. *A worried homeowner hears of your computations and asks, 'How cold would it have to get to contract my house's volume by 1%? I don't want to break the china.' What is your answer?*

7.1. Volume Expansion Explained by Calculus

Once you are comfortable with the symbolic rules of calculus, the large change in volume is easily understood. The instantaneous rate of thermal expansion of the length x of an object is the derivative $\frac{dx}{dT}$. An object originally of unit length therefore has $\frac{dx}{dT} = c$. Our house has dimensions $l = 40 x$, $w = 30 x$, and $h = 8 x$ where the initial length $x_0 = 1$ foot. In terms of x, the volume of the house is

$$V = l \, w \, h = (40 \, x)(30 \, x)(8 \, x) = 9600 \, x^3$$

The chain rule says

$$\frac{dV}{dT} = \frac{dV}{dx} \frac{dx}{dT}$$

and basic rules give

$$\frac{dV}{dx} = 9600 \cdot 3 \, x^2 = 3 \, V_0 \, x^2, \quad \text{so} \quad \frac{dV}{dT} = 3 \, V_0 \, x^2 \frac{dx}{dT}$$

At the initial temperature $x = 1$, thus we have $\frac{dV}{dT} = 3 \, V_0 \, c$ or, in terms of differentials, the change in V is given by

$$dV = 3 \, V_0 \, c \, dT$$

Numerically, $3 \, V_0 \, c = .115$, so that a change of temperature of $dT = 16$ produces an approximate change in volume of

$$dV = .115 \times 16.0 = 1.84 \quad \text{cu. ft.}$$

The insight into the large volume expansion is clearest in the initial symbolic formula for expansion

$$dV = 3 \, V_0 \, c \, dT$$

because the triple volume compensates for the small linear expansion coefficient, c.

HINT 7.9. *Relative Expansion*
While the linear expansion is small on our familiar scale, the volume seems larger - a bathtub full. However, it might be more reasonable to compare the change in volume to the amount we started with. Write an expression for the percentage change in volume and solve for that in terms of c and the change in temperature. Show that the percent change in volume is three times the percent change in length.

EXPLORATION 7.10. *Do the scientific tables giving thermal expansion per unit length mean that*

$$\frac{dx}{dT} = c\,x_0 \quad or \quad \frac{dx}{dT} = c\,x$$

where x_0 is the fixed length at the initial temperature, while x is the varying length as temperature changes?

The first differential equation describes a straight line of constant slope $x_0 c$, so the solution is $x = x_0(1 + cT)$. We will see in Chapter 8 of the core text that the solution to the second equation is $x = x_0\, e^{cT}$. Use Mathematica to plot both functions over the range $-50 \leq T \leq 50$ with $x_0 = 1$ and $c = 4.0 \times 10^{-6}$. What do you observe? In particular, what is the largest difference between $(1 + cT)$ and e^{cT} over this range?

Chapter 4

Applications of the Increment Approximation
$$f[t + \delta t] - f[t] = f'[t]\,\delta t + \varepsilon \cdot \delta t$$

PROJECT 8

A Derivation of Hubble's law

Evidence of an expanding universe is one of the most important astronomical observations of this century. Light received from a distant galaxy is "old" light, generated years ago at a time t_e when it was emitted. When this old light is compared to light generated at the time received t_r, it is found that the characteristic colors, or spectral lines, do not have the same wavelengths.

The red shift was discovered by V. M. Slipher (1875 — 1969), who measured shifts for more than 20 galaxies during 1912 — 1925. The observed wavelengths from all distant galaxies are longer than the emitted wavelengths (assuming constant spectra), $\lambda_r > \lambda_e$. This is called the "red shift," because red light has longer wavelengths than blue light.

The Doppler effect says that wavelengths from a transmitter that is moving away from us are longer. The Doppler effect is familiar from the sound of a speeding truck approaching, passing and receding, eeeeeee – aaaaaaa. (Lower pitch means longer wavelength for the receding truck.) The Doppler shift is connected with the velocity of the emitter, v_e, by the formula

$$\frac{\lambda_e}{\lambda_r} = \frac{1}{1 + v_e/c}$$

where c is the speed of light.

Since all distant galaxies are red-shifted, they all are moving away from us. It seems they can all be moving away only if the universe is expanding. (If we were simply moving through the universe, we would be moving away from some that would appear red-shifted and toward others that would appear blue-shifted.)

In 1929 E. P. Hubble (1889 — 1953) announced the empirical rule now called Hubble's law:

$$\frac{\lambda_e}{\lambda_r} = 1 - const \cdot D$$

where D is the distance the light traveled from the galaxy. The constant is now written in terms of the speed of light c and a constant H named for Hubble:

$$\frac{\lambda_e}{\lambda_r} = 1 - \frac{H}{c} \cdot D$$

Older galaxies with large difference in time are more distant, because light travels at constant speed and we observe them all in our relatively short lifetime. Suppose we let t denote the time (in years, for example), with some specific time set as $t = 0$. The time that the light is received t_r is fixed at our lifetime, but the time that the light was emitted t_e varies from galaxy to galaxy (and $t_e < t_r$.) The age of the light equals the difference $t_r - t_e$ and is proportional to D, because distance equals the rate times the time (in the appropriate units.)

HINT 8.1. *Write the age of the light in terms of D and the speed of light c, $t_r - t_e = ?$*

Hubble's law means that the old expansion is faster than the new expansion. Explain why this is so. Notice that older light travels a longer distance D and produces a smaller ratio.

The math here is simple, but the scientific interpretation takes some thought. It might help to rewrite Hubble's law by replacing D/c by the amount of time the light has traveled:

$$\frac{\lambda_e}{\lambda_r} = 1 - H \cdot (??)$$

Faster expansion means that the wavelengths are shifted more, so that λ_e/λ_r is closer to zero. You want to show that if the light is very old ($t_r - t_e$ is large), then λ_e is "very shifted" from λ_r (so λ_e is small compared to λ_r), whereas if the light is not so old, then λ_e is not so shifted from λ_r.

The next problem uses the microscope approximation two ways and the continuity approximation of functions to show that Hubble's law is a "local" consequence of Doppler's law given above.

EXPLORATION 8.2. *Show that Hubble's law follows from the "microscope" approximation by working through the following steps. We will use the function $D(t)$ for the distance from earth to the galaxy emitting light as a function of time.*

(1) *Show how to use the microscope approximation*

$$f[x + \delta x] = f[x] + f'[x] \cdot \delta x + \varepsilon \cdot \delta x$$

for the function $D(t)$ to give the two approximations

$$D[t_r] \approx D[t_e] + D'[t_e] \cdot (t_r - t_e) \quad \text{and} \quad D[t_e] \approx D[t_r] - D'[t_r] \cdot (t_r - t_e)$$

(Hints: First use $x = t_e$ and $\delta x = (t_r - t_e)$, then use $x = t_r$. Be careful with signs.)

(2) *All light travels at the speed c, so you can express $\Delta t = (t_r - t_e)$ in terms of the distance $D[t_e]$ and the speed of light, c. Do this and substitute the expression in the above approximations.*

(3) *Divide both sides of the second approximation $D[t_e] \approx D[t_r] - D'[t_r] \cdot (t_r - t_e)$ by $D[t_r]$ to show that*

$$\frac{D[t_e]}{D[t_r]} \approx 1 - \frac{D'[t_r]}{D[t_r]} \cdot \frac{D[t_e]}{c}$$

(4) *Hubble's constant is defined as "the ratio of the speed at which a distant galaxy is receding from the earth to its distance from the earth." This is the ratio*

$$\frac{D'[t_e]}{D[t_e]}$$

and is currently estimated somewhere between 20 and 100 kilometers per second per million parsecs. A parsec is an astronomical unit equal to 3.26 light-years or 3.09×10^{13} km. (The speed of light in vacuum is 2.997925×10^5 km/sec, so light travels 9.46×10^{12} km in one year.)

Consider the term
$$\frac{D'[t_r]}{D[t_r]}$$
in the approximation of (3) for $D[t_e]/D[t_r]$. If $t_r - t_e$ is small, we have
$$\frac{D'[t_r]}{D[t_r]} \approx \frac{D'[t_e]}{D[t_e]}$$
Which theorems in the main text justify this?

(5) Let $H = D'[t_r]/D[t_r] \approx D'[t_e]/D[t_e]$ and substitute into the approximation of (3), to give the approximation
$$\frac{D[t_e]}{D[t_r]} \approx 1 - \frac{H}{c} \cdot D$$
but not Hubble's law.

(6) Use the first approximation of (1) above, $D[t_r] \approx D[t_e] + D'[t_e] \cdot (t_r - t_e)$, to show that
$$\frac{D[t_e]}{D[t_r]} \approx \frac{D[t_e]}{D[t_e] + D'[t_e] \cdot (t_r - t_e)}$$
Do some additional substitution to see that
$$\frac{D[t_e]}{D[t_r]} \approx \frac{1}{1 + D'[t_e]/c}$$
The velocity of the galaxy at the time the light was emitted is $D'[t_e]$, so the expression above yields
$$\frac{D[t_e]}{D[t_r]} \approx \frac{1}{1 + v_e/c}$$
where v_e is the velocity of the emitter and c is the speed of light.

(7) Show that this gives Hubble's law in terms of the lambdas, H, D, and c by using the two expressions for $D[t_e]/D[t_r]$ and the Doppler law (at the beginning of the section) for the ratio of the lambdas.

Contemporary astronomers no longer believe that Hubble's constant is actually constant. Why would Hubble's law still appear to be true in (astronomically) short periods of time?

PROJECT 9

Functional Linearity

Linearity in function notation has a peculiar appearance. This is not difficult, just different. The exercises of this section give the linear case of the main formula underlying differential calculus,

$$f[x + \delta x] - f[x] = f'[x] \cdot \delta x + \varepsilon \cdot \delta x$$

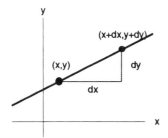

FIGURE 9.1: Linear Change in y

We want to consider the linear equation as a function, $f[x] = m\,x + b$, and compare perturbed values of the function symbolically and graphically. Some numerical computations may help get you started, but you should try to understand the fundamental role of the parameters m and b in general.

HINT 9.2. *Functional Linearity*

(1) *Choose convenient values of the parameters m and b and make a sketch of the graph of your function $y = f[x]$. Do this for several cases of m and b.*
(2) *Fix one value of x and mark the (x, y) point on your graphs over this x.*
(3) *Choose a perturbation Δx and mark the point above $x + \Delta x$ on your graphs. Each of your graphs should look something like Figure 9.1.*
(4) *Where does the value $y_1 = f[x]$ appear on your graph?*
(5) *Where does the value $y_2 = f[x + \Delta x]$ appear on your graph?*

(6) How much do you need to add to the vertical value y_1 to move to the vertical value y_2?
   i) Express your answer in terms of m, b, x, and Δx.
   ii) Express your answer in terms of $f[x]$ and $f[x + \Delta x]$.
(7) Substitute $x + \Delta x$ into the formula for $f[\cdot]$ and show symbolically that
$$f[x + \Delta x] - f[x] = m \, \Delta x$$
In other words, no matter which x we begin at, the size of the change in y caused by a change of Δx in x is $m \, \Delta x$ and does not depend on b or x. This formula could be expressed
$$f[x + \Delta x] - f[x] = m \cdot \Delta x$$
where m is constant. (Recall that m is a parameter, so does not depend on either x or Δx.)

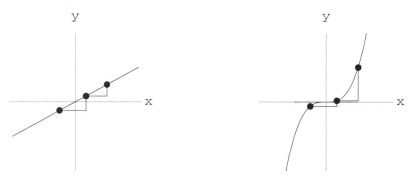

FIGURE 9.3: Linear and Nonlinear Changes in y

The important formula of the previous exercise is not true for nonlinear functions. It fails in two ways. Consider $f[x] = a \, x^2$, for example. If I say, "add 1 to x and tell me how much f changes," you can only answer in terms of x.

$$f[x + 1] - f[x] = a \, (x + 1)^2 - a \, x^2 = a \, ((x + 1)^2 - x^2)$$
$$= a \, (x^2 + 2 \, x + 1 - x^2) = a \, (2 \, x + 1)$$

If f were linear, $f[x] = m \, x + b$, the answer would be independent of x.

$$f[x + 1] - f[x] = (m \, (x + 1) + b) - (m \, x + b) = m$$

HINT 9.4. *Functional Nonlinearity*
Let $f[x] = a \, x^2$. Compute $f[x + \Delta x] - f[x]$ symbolically and show that when we write the change in the function
$$f[x + \Delta x] - f[x] = g[x, \Delta x] \cdot \Delta x$$
for a function $g[x]$ that depends both on x and on Δx. (Hint: Compute $(f[x + \Delta x] - f[x])/\Delta x$.)

The function $f[x] = a \, x^2$ is smooth, so it satisfies the increment approximation
$$f[x + \delta x] - f[x] = f'[x] \cdot \delta x + \varepsilon \cdot \delta x$$
where $\varepsilon \approx 0$ when $\delta x \approx 0$ and $f'[x] = 2 \, a \, x$.

HINT 9.5. *Compare the two expressions for the change in $f[x]$, your function*
$$g[x, \delta x] = f'[x] + \varepsilon[x, \delta x]$$
when x is fixed and $\Delta x - \delta x \approx 0$ is small. Use your formula to explain: Why does the graph of $y = a x^2$ appear linear on a small scale? Why does the slope of a microscopic view of the graph change when we chnage x?

PROJECT 10

Functional Identities

The use of "unknown functions" is of fundamental importance in calculus and other branches of mathematics and science. We shall often see in the course that differential equations can be viewed as identities for unknown functions. One reason that students sometimes have difficulty understanding the meaning of derivatives or even the general rules for finding derivatives is that those things involve equations in unknown functions. The symbolic rules for differentiation and the increment approximation defining derivatives are similar to general functional identities in that they involve an unknown function. It is important for you to get used to this 'higher type variable,' an unknown function. This chapter can form a bridge between the specific identities of high school and the unknown function variables from rules of calculus and differential equations.

In high school you learned that trig functions satisfy certain identities or that logarithms have certain "properties." All the the identities you need to recall from high school are

$$(\text{Cos}[x])^2 + (\text{Sin}[x])^2 = 1 \qquad \text{CircleIden}$$
$$\text{Cos}[x+y] = \text{Cos}[x]\,\text{Cos}[y] - \text{Sin}[x]\,\text{Sin}[y] \qquad \text{CosSum}$$
$$\text{Sin}[x+y] = \text{Sin}[x]\,\text{Cos}[y] + \text{Sin}[y]\,\text{Cos}[x] \qquad \text{SinSum}$$
$$b^{x+y} = b^x\, b^y \qquad \text{ExpSum}$$
$$(b^x)^y = b^{x \cdot y} \qquad \text{RepeatedExp}$$
$$\text{Log}[x \cdot y] = \text{Log}[x] + \text{Log}[y] \qquad \text{LogProd}$$
$$\text{Log}[x^p] = p\,\text{Log}[x] \qquad \text{LogPower}$$

but you must be able to *use* these identities. Some practice exercises using these familiar identities are given in the text CD Chapter 28 on high school review. The Math Background book has more information.

A general *functional identity* is an equation that is satisfied by an unknown function (or a number of functions) over its domain. For example, the function

$$f[x] = 2^x$$

satisfies $f[x+y] = 2^{(x+y)} = 2^x 2^y = f[x]f[y]$, so eliminating the two middle terms, we see that the function $f[x] = 2^x$ satisfies the functional identity

(ExpSum) $$f[x+y] = f[x] \cdot f[y]$$

It is important to pay attention to the variable or variables in a functional identity. In order for an equation involving a function to be a functional identity, the equation must be valid for *all* values of the variables in question. Equation (ExpSum) is satisfied by the function $f[x] = 2^x$ for all x and y. For the function $f[x] = x$, it is true that $f[2+2] = f[2]f[2]$, but $f[3+1] \neq f[3]f[1]$, so $f[x] = x$ does not satisfy functional identity (ExpSum).

HINT 10.1. *Verify that for any positive number, b, the function $f[x] = b^x$ satisfies the functional identity (ExpSum) for all x and y. Is (ExpSum) valid (for all x and y) for the function $f[x] = x^2$ or $f[x] = x^3$? Justify your answer.*

HINT 10.2. *Define $f[x] = \text{Log}[x]$ where x is any positive number. Why does this $f[x]$ satisfy the functional identities*

(LogProd) $$f[x \cdot y] = f[x] + f[y]$$

and

(LogPower) $$f[x^k] = k \cdot f[x]$$

where x, y, and k are variables. What restrictions should be placed on x and y for the equations to be valid? What is the domain of the logarithm?

Functional identities are a sort of 'higher laws of algebra.' Observe the notational similarity between the distributive law for multiplication over addition,

$$m \times [x+y] = m \times x + m \times y$$

and the additive functional identity

(Additive) $$f[x+y] = f[x] + f[y]$$

Most functions $f[x]$ do not satisfy the additive identity, for example,

$$\frac{1}{x+y} \neq \frac{1}{x} + \frac{1}{y} \quad \text{and} \quad \sqrt{x+y} \neq \sqrt{x} + \sqrt{y}$$

The fact that these are not identities means that for *some* choices of x and y in the domains of the respective functions $f[x] = 1/x$ and $f[x] = \sqrt{x}$, the two sides are not equal. Using the Mathematical Background book, Chapter 2, you can show that the only differentiable functions that do satisfy the additive functional identity are the functions $f[x] = m \cdot x$. In other words, the additive functional identity is nearly equivalent to the distributive law; the *only* unknown (differentiable) function that satisfies it *is* multiplication. Other functional identities such as the seven given at the start of this chapter capture the most important features of the functions that satisfy the respective identities. For example, the pair of functions $f[x] = 1/x$ and $g(x) = \sqrt{x}$ do not satisfy the addition formula for the sine function, either.

HINT 10.3. *Find values of x and y so that the left and right sides of each of the additive formulas for $1/x$ and $\sqrt{x}$ are not equal.*

Show that $1/x$ and $\sqrt{x}$ also do not satisfy the identity (SinSum), that is,

$$\frac{1}{x+y} = \frac{1}{x}\sqrt{y} + \sqrt{x}\,\frac{1}{y}$$

is false for some choices of x and y in the domains of these functions.

The goal of this chapter is to give you practice at working with familiar functional identities and to extend your thinking to identities in unknown functions.

HINT 10.4. *1) Suppose that $f[x]$ is an unknown function that is known to satisfy (LogProd) (so $f[x]$ behaves "like" $\mathrm{Log}[x]$, but we don't know if $f[x]$ is $\mathrm{Log}[x]$), and suppose that $f[0]$ is a well-defined number (even though we don't specify exactly what $f[0]$ is). Show that this function $f[x]$ must be the zero function, that is, show that $f[x] = 0$ for every x. (Hint: Use the fact that $0 * x = 0$.)*

2) Suppose that $f[x]$ is an unknown function that is known to satisfy (LogPower) for all $x > 0$ and all k. Show that $f[1]$ must equal 0, $f[1] = 0$. (Hint: Fix $x = 1$, and try different values of k.)

HINT 10.5. *Let m and b be fixed numbers and define*

$$f[x] = mx + b$$

Verify that if $b = 0$, this function satisfies the functional identity

(Mult) $\qquad f[x] = x\, f[1]$

for all x and that if $b \neq 0$, $f[x]$ will not satisfy (Mult) for all x (that is, given a nonzero b, there will be at least one x for which (Mult) is not true).

Prove that any function satisfying (Mult) also automatically satisfies the two functional identities

(Additive) $\qquad f[x+y] = f[x] + f[y]$

and

(Multiplicative) $\qquad f[x\,y] = x f[y]$

for all x and y.

Suppose $f[x]$ is a function that satisfies (Mult) (and for now that is the only thing you know about $f[x]$). Prove that $f[x]$ must be of the form $f[x] = m \cdot x$, for some fixed number m (this is almost obvious).

Prove that a general power function $f[x] = mx^k$, where k is a positive integer and m is a fixed number, will not satisfy (Mult) for all x if $k \neq 1$ (that is, if $k \neq 1$, there will be at least one x for which (Mult) is not true).

Prove that $f[x] = \mathrm{Sin}[x]$ does not satisfy the additive identity.

Prove that $f[x] = 2^x$ does not satisfy the additive identity.

53

10.1. Additive Functions

In the early 1800s, Cauchy asked the question: Must a function satisfying

(Additive) $$f[x+y] = f[x] + f[y]$$

be of the form $f[x] = m \cdot x$? This was not solved until the late 1800s by Hamel. The answer is "No." There are some very strange functions satisfying the additive identity that are not simple linear functions. However, these strange functions are not differentiable.

Chapter 2, Functional Identities from the Mathematical Background book, Foundations of Infinitesimal Calculus, explores this problem in more detail.

Chapter 5

Differential Equations from Increment Geometry

The three curves in this chapter illustrate an important idea of calculus: they are each described by a differential equation, which in turn comes from studying properties of a tiny increment of the curve without knowing the curve itself in advance. The curves have a physical description – they are quite concrete – but the tractrix and catenary provide important examples in differential geometry and the calculus of variations. In other words, these curves are important in mathematics.

PROJECT 11

The Tractrix

It's spring and you just finished a calculus exam. You leave the exam room with your book tied to a rope 1 m long. When you reach the sidewalk, you toss the book straight out perpendicular to the walk into the mud. Then you proceed down the sidewalk holding the rope and making a trail in the mud with your book. The tractrix is the path you've made in the mud with your human-powered tractor. A rough figure is Figure 1.1.

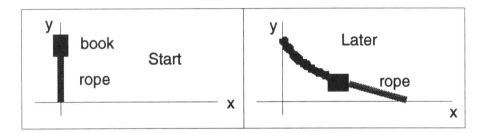

FIGURE 11.1: Tractrix Sketch

Mathematically, the length of the rope remains 1 m. and it acts to pull the book tangent to the furrow in the mud. We will use standard (x,y) coordinates to express these two facts. If the book is at the point (x,y) and you are at the point $(x+h,0)$ on the sidewalk, then the length of the rope is the hypotenuse of a right triangle with base h and height y, so the Pythagorean theorem says

$$1 = h^2 + y^2$$

HINT 11.2. *Draw the triangle from the position of the book at (x,y) perpendicular to the x-axis at $(x,0)$ and over to your position at $(x+h,0)$. Verify that this tells us that*

$$h = \sqrt{1 - y^2}$$

The slope of the furrow is $\frac{dy}{dx}$ and that must also be equal to the slope of the hypotenuse of the triangle you just drew in the previous exercise.

HINT 11.3. *Show that the slope of the hypotenuse of your triangle is*

$$-\frac{y}{h} = \frac{-y}{\sqrt{1-y^2}}$$

Equating the two slopes gives us a differential equation for the furrow, a curve known as the tractrix,

$$\frac{dy}{dx} = \frac{-y}{\sqrt{1-y^2}}$$

HINT 11.4. *Separate variables in the equation above for the tractrix and integrate both sides. (Use Mathematica if you wish, but be warned that we need the range where y starts at 1 and decreases.)*

$$-\frac{\sqrt{1-y^2}}{y} dy = dx$$

$$\int \frac{\sqrt{1-y^2}}{y} dy = -x + C$$

$$\int \frac{\sqrt{1-(\cos[\theta])^2}}{\cos[\theta]} (-\sin[\theta]) \, d\theta = -x + C$$

$$\text{using } y = \cos[\theta], \quad dy = (-\sin[\theta]) \, d\theta$$

$$\vdots$$

$$x = C + \int \frac{1 - \cos^2[\theta]}{\cos[\theta]} d\theta$$

$$x = C + ???$$

Since $x = 0$ at the start when $y = 1$, the value of your constant of integration C is ???

Generalize this derivation of the tractrix to a rope of length k, showing that

$$\frac{dy}{dx} = \frac{-y}{\sqrt{k^2 - y^2}}$$

$$x = k \, \text{Log} \left[\frac{k + \sqrt{k^2 - y^2}}{y} \right] - \sqrt{k^2 - y^2}$$

The formula for x as a function of y is not really what we would like in terms of the physical question. We would like to specify the point $(x + h, 0)$ and compute both x and y. The differential equation needs to be used directly for those sorts of computations.

PROJECT 12

The Isochrone

A bead slides along a frictionless wire and starts with an initial downward velocity v_0. Find the shape of the wire that maintains a constant downward component of velocity. In other words, find the shape that converts all the gravitational energy into new horizontal kinetic energy. Rather than an energy argument, we will begin by finding equations of motion using Newton's law, $F = mA$, or total applied force equals mass times the second derivative of position.

Since the wire is frictionless, it can only produce a force perpendicular to the bead. Since the vertical speed cannot increase, the weight of the bead, mg must equal the upward portion of the wire force. This is shown in the rough sketch Figure 12.1.

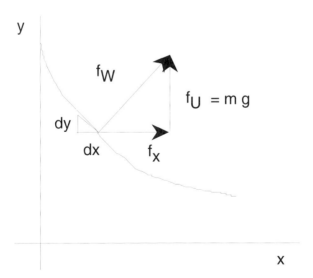

FIGURE 12.1: Isochrone Forces

The fact that the force from the wire is perpendicular to the wire means that a tiny

incremental triangle along the wire is similar to the triangle of forces, yielding
$$\frac{dy}{dx} = \frac{f_x}{f_U} = \frac{f_x}{m\,g}$$

We may use the Chain Rule to express the geometric slope in terms of the horizontal and vertical speeds,
$$\frac{dy}{dx} = \frac{dy}{dt} \bigg/ \frac{dx}{dt} = \frac{f_x}{m\,g}$$

The vertical speed, $\frac{dy}{dt} = v_0$, is to remain constant (since we balanced the force of gravity with the upward component of wire force), so
$$v_0 \bigg/ \frac{dx}{dt} = \frac{f_x}{m\,g}$$
$$\frac{dx}{dt} = \frac{m\,g\,v_0}{f_x}$$
$$f_x \frac{dx}{dt} = m\,g\,v_0$$

The horizontal motion is goverened by Newton's $F = mA$ law,
$$f_x = m \frac{d^2 x}{dt^2}$$

and we combine this with the previous equation to obtain
$$\frac{dx}{dt} \frac{d^2 x}{dt^2} = g\,v_0$$

Now use the phase variable trick; let $u = \frac{dx}{dt}$ so the equation above becomes
$$u \frac{du}{dt} = g\,v_0$$

HINT 12.2. *Separate variables in the horizontal speed equation above and show that*
$$u = \sqrt{2\,g\,v_0\,t}$$

if we suppose that the initial horizontal velocity is zero, $u(0) = 0$.

Use the equation for $u = \frac{dx}{dt}$ to show that the horizontal position is given by
$$x = \frac{\sqrt{g\,v_0}}{3} [2\,t]^{3/2}$$

for $x(0) = 0$.

We also know (if we measure y downward, so its velocity is positive) that $\frac{dy}{dt} = v_0$.

HINT 12.3. *Prove that the vertical position is given by*
$$y = v_0\,t$$

if we start at $y(0) = 0$.

We can solve the y equation fot $t = y/v_0$ and substitute into the horizontal position equation, obtaining $x = \frac{\sqrt{g\,v_0}}{3} [2\,\frac{y}{v_0}]^{3/2}$.

HINT 12.4. *Solve this equation for $y = y(x) = k\ x^{2/3}$ showing that $y = \frac{1}{2} \sqrt[3]{\frac{(3\,v_0)^2}{g}}\ x^{2/3}$.*

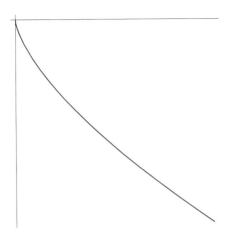

FIGURE 12.5: The Isochrone

We have found several equations for the isochrone.

Parametric $\qquad\qquad x = x(y) \qquad\qquad y = y(x)$

$$x(t) = \frac{\sqrt{g\,v_0}}{3}\,[2\,t]^{3/2} \qquad x = \frac{\sqrt{g\,v_0}}{3}\,[2\,\frac{y}{v_0}]^{3/2} \qquad y = \frac{1}{2}\sqrt[3]{\frac{(3\,v_0)^2}{g}}\,x^{2/3}$$

$$y(t) = v_0\,t$$

12.1. Conservation of Energy

The equations for the isochrone can also be found from an energy argument. You need to know that the increase in kinetic energy changing from $\frac{dx}{dt}(0) = 0$ to a higher horizontal speed u is

$$\frac{1}{2}\,m\,u^2$$

You also need to know that the decrease in gravitational potential energy in moving down a distance y is

$$m\,g\,y$$

The physical principle of conservation of energy (which we showed for a linear oscillator in Chapter 17 of the main text) says these two quantities are equal, so

$$g\,y = \frac{1}{2}\left(\frac{dx}{dt}\right)^2$$

HINT 12.6. *Use the Chain Rule to express the differential equation above in the form*

$$y = \frac{v_0^2}{2\,g}\left(\frac{dx}{dy}\right)^2$$

Then separate variables and show that

$$x = \frac{\sqrt{g\, v_0}}{3} [2\frac{y}{v_0}]^{3/2}$$

as before.

How could you find the equation for $x(t)$ from this equation?

PROJECT 13

The Catenary

We want to find the shape and length of electrical wires hanging between telephone poles. Intuitively, we know that we cannot stretch the wire perfectly straight, because there has to be some upward component of the tension force to counteract the weight of the wire and hold it up. The total length is one component of the cost in hanging such wires, but the strength of the wire is a counterbalancing cost. If we want to make the wire very nearly straight, then we need to make it very strong to support the tension required in straightening it. We will develop a model that has an optimal shape (or amount of sag) in terms of the tension (that is, minimal required strength). Our model is based on two simplifying assumptions:

13.1. The Catenary Hypotheses

(1) The wire is perfectly flexible, that is, supports no bending forces.
(2) The wire is perfectly inelastic, that is, does not stretch.

Catena in Latin means "chain." A chain is very strong, so it does not stretch very much, and because of the links, it does not support any bending force beyond a single link. We may imagine that our model curve is a chain made of infinitely strong infinitesimal links. The center line of this chain is the curve we seek to describe.

There are two parameters in our model, the linear density of the chain and the horizontal force at the pole. Imagine hanging the chain by hanging it over the poles and then pulling on it to shorten the sag. The pulling force is H. The weight of the chain is a function of its length.

13.2. Parameters

(1) w – the linear weight density (n/m) of the chain.
(2) H – the horizontal tension in the chain at the pole in newtons.

The m-k-s unit of force is "newtons." A copper wire 1 cm in diameter weighs about 5 n/m. (A mass of 1 kilogram weighs $g = 9.8$ n.) Linear weight density means that a chain L meters long weighs $w \cdot L$ newtons.

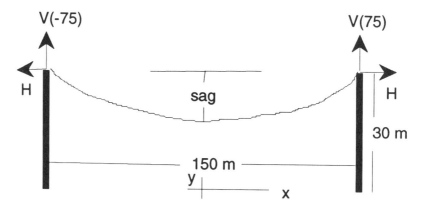

FIGURE 13.1: A Catenary Suspended over 30-m Poles 150 m Apart

We place an (x, y)-coordinate system on the ground below the low point of the catenary.

13.3. Variables

(1) x – the distance measured horizontally to a point on the catenary (m)
(2) y – the distance measured vertically from the ground to a point on the catenary (m)
(3) L – the length along the catenary from 0 to x (m)
(4) **T** – the total tension vector above x (n)
(5) V – the horizontal component of tension above x, $\mathbf{T}(x) = (H, V(x))$ (n)

13.4. The Equation for Tension

A hanging electrical wire remains at rest, so its velocity and acceleration are zero. Newton's $F = m A$ law implies that the total force must also be zero, or that the tension forces and gravity must sum to zero. A rough sketch of a segment of wire is shown in Figure 13.2. We imagine cutting out a segment and holding the ends at the tension that would normally be present keeping it stationary.

The fact that the catenary does not move left or right means that the horizontal acceleration is zero and the horizontal forces cancel, so H is constant or $H(x) = H(x + dx)$, for any size x and dx.

There are three vertical forces on a piece of the catenary, the upward pull at $x + dx$, the downward pull at x and the weight $w \cdot [L(x+dx) - L(x)]$ acting down. "No vertical motion," means that the total force up must equal the total force down, or

$$V(x + dx) = V(x) + w \cdot [L(x + dx) - L(x)]$$

for any size x and dx along the catenary. This functional identity may also be written $V(x + dx) - V(x) = w \cdot [L(x + dx) - L(x)]$, the change in vertical force equals the change in weight.

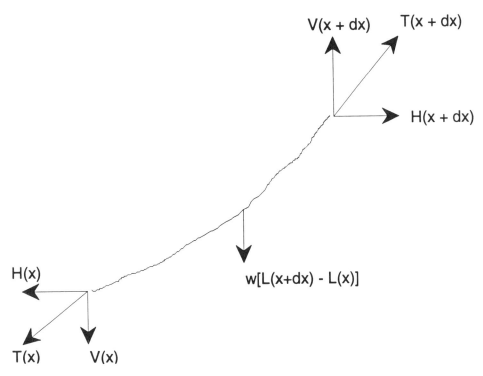

FIGURE 13.2: Tension on a Small Segment

To find a differential equation for the vertical tension, we will examine a small increment of the catenary. If the catenary is a differentiable curve, the Increment Principle (see Chapters 3, 4, 5, and 18 of the main text) says that the magnified curve will appear to be a straight line, so if we mark changes in x and y on the microscopic view, we will see a traingle. This triangle tells us two things; (1) the approximate length of an increment and (2) the relation between the components of tension and the slope.

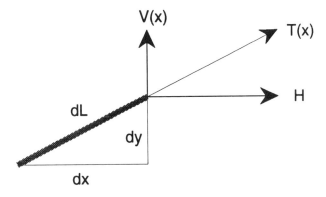

FIGURE 13.3: A Tiny Increment of the Catenary

The length of the portion of the catenary is then given by the Pythagorean theorem, since the catenary forms the hypotenuse of the apparent triangle.

$$L(x + \delta x) - L(x) \approx\approx \sqrt{\delta x^2 + \delta y^2}$$

Actually, we should be a little more precise about the approximation. What we know is that the error is small on the scale of the microscope, or

$$\delta L = \sqrt{\delta x^2 + \delta y^2} + \varepsilon \cdot \delta x$$

with $\varepsilon \approx 0$.

Since the catenary cannot support any bending force, the tension force must act tangent along the catenary or the finite size horizontal and vertical components of tension must form a triangle similar to the increment triangle. In other words, the slope of the curve must equal the slope of the tension triangle,

$$\frac{\delta y}{\delta x} \approx \frac{dy}{dx} = \frac{V(x)}{H}$$

The equation that says there is no vertical motion applies to the small increment (by the Function Extension Axiom) to tell us

$$V(x + dx) - V(x) = w \cdot [L(x + dx) - L(x)]$$
$$V(x + \delta x) - V(x) = w \cdot [L(x + \delta x) - L(x)]$$
$$\delta V = w \cdot \delta L$$
$$\delta V = w \cdot [\sqrt{\delta x^2 + \delta y^2} + \varepsilon \cdot \delta x]$$
$$\frac{\delta V}{\delta x} = w \cdot [\sqrt{\frac{\delta x^2}{\delta x^2} + \frac{\delta y^2}{\delta x^2}} + \varepsilon]$$
$$\frac{\delta V}{\delta x} \approx w \cdot \sqrt{1 + \frac{[V(x)]^2}{H^2}}$$

HINT 13.4. *The Vertical Tension*
Show that the vertical component of the tension in a catenary satisfies the initial value problem

$$V(0) = 0$$
$$\frac{dV}{dx} = w \cdot \sqrt{1 + \frac{[V(x)]^2}{H^2}}$$

Why is the vertical tension zero at the bottom of the catenary?
Show that the unique solution of this problem is

$$V(x) = H \; Sinh\left[\frac{w \; x}{H}\right]$$

In other words, show that substitution of this function satisfies the differential equation and initial condition. Does this initial value problem satisfy the hypotheses of Theorem 17.20 of the main text?

See Exercise 18.33 of the main text for more on "hyperbolic trig functions." The basic facts are

$$\text{Sinh}[u] = \frac{e^u - e^{-u}}{2} \qquad\qquad \text{Cosh}[u] = \frac{e^u + e^{-u}}{2}$$

$$\frac{d\text{Sinh}[u]}{du} = \text{Cosh}[u] \qquad\qquad \frac{d\text{Cosh}[u]}{du} = \text{Sinh}[u]$$

$$\text{Cosh}^2[u] - \text{Sinh}^2[u] = 1$$

HINT 13.5. *The Position*
Show that the differential equation for $y = y(x)$ *along the catenary is*

$$\frac{dy}{dx} = \text{Sinh}\left[\frac{w\,x}{H}\right]$$

so that, we can write $y(x)$ *in the form*

$$y = \frac{H}{w}\left(\text{Cosh}\left[\frac{w\,x}{H}\right] - 1\right) + Y_0$$

where $y(0) = Y_0$. *(Hint: How much are* $\text{Cosh}[0]$ *and* $\frac{dy}{dx}$*?)*
Use Mathematica to plot $y(x)$ *when* $w = 5$, $H = 1000$, *and* $Y_0 = 20$.

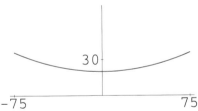

FIGURE 13.6: Hyperbolic Cosine

There is a problem with the expression that we have given for the position of the catenary. We want to have $y(\pm 75) = 30$, so that the wire is at the top of the poles when $x = \pm 75$. We also know that we physically adjust the amount of sag in the catenary by pulling on the chain or changing H. In other words, we do not know Y_0, but set it by finding H.

HINT 13.7. *The Sag*
Show that the sag in the catenary satisfies the equation

$$sag = \frac{H}{w}\left(\text{Cosh}\left[\frac{w\,75}{H}\right] - 1\right)$$

(Hint: How much is $y(75) - y(0)$ *even if you do not know* $y(0)$*?)*
  Given a catenary with $w = 5$, *find the horizontal tension* H *needed to produce a sag of 10 m. Use Mathematica's root finding command,*
    FindRoot[y == s,{h,1000}]
FindRoot[] approximates a root to the equation $y = s$ *given an initial guess of* $h = 1000$. *Once you have found the correct tension, find the equation for the catenary and Plot your result. (As a check, how much is* $y(0)$*?)*

The strength needed in the catenary is governed not just by the vertical component of the tension but rather by the magnitude of the whole tension vector.

HINT 13.8. *The Highest Tension*
Show that the magnitude of the tension vector at any point x along the catenary is given by

$$|\mathbf{T}(x)| = H \; Cosh\left[\frac{w\,x}{H}\right]$$

(Hint use the identity $Cosh^2[u] - Sinh^2[u] = 1$ and the two components of tension.)
Prove that the maximum tension occurs at the endpoint,

$$M = Max[|\mathbf{T}(x)| : 0 \leq x \leq r] = H \; Cosh\left[\frac{w\,r}{H}\right]$$

and use this to find the maximal tension in the catenary with $w = 5$ and a sag of 10 meters in 75.

Now that we know the strength required to make a catenary with a sag of 10 m in 75, we want to find its length. We saw above in this section (and in Chapter 14 of the main text) that the length of a tiny piece of the catenary is

$$\delta L = \sqrt{\delta x^2 + \delta y^2} + \varepsilon \cdot \delta x$$

$$= \sqrt{1 + \left(\frac{\delta y}{\delta x}\right)^2}\, \delta x + \varepsilon \cdot \delta x$$

Hence, the length of the catenary is given by

$$L(75) - L(0) = \int_0^{75} \sqrt{1 + \left(\frac{dy}{dx}\right)^2}\, dx$$

HINT 13.9. *The Length*
Show that the length of the catenary from the low point at $x = 0$ to the pole at $x = 75$ is

$$Length = \frac{H}{w}\, Sinh\left[\frac{w\,75}{H}\right]$$

(A shortcut to integration would be to compare the total weight to the upward force at the pole.) Use this to calculate the length of a catenary with $w = 5$ and a 10-m sag in 75 meters.

13.5. Optimizing Length and Strength

Intuitively, we know that we cannot stretch the catenary perfectly straight between the poles. Verify this mathematically:

HINT 13.10. *The Limit as $H \to \infty$ or sag$\to 0$*
Prove that it requires huge maximum total tension, $|\mathbf{T}(75)|$, to make a catenary with a tiny sag, or if sag≈ 0, then H must be infinite. (Hint: Use the equation relating H to sag.)

The previous exercise shows that some sag is desirable for chains of finite strength, but it should also be clear that the sag should not be too big. In fact, even neglecting the fact that the catenary would touch the ground unless the poles were made taller, a very long cable produces very high weight.

HINT 13.11. *The Limit as $H \to 0$ or sag$\to \infty$*
Prove that it requires huge maximum total tension, $M = |\mathbf{T}(75)|$, to make a catenary with a huge sag, or if sag is infinite, then $M = |\mathbf{T}(75)|$ must be infinite. (Hint: Use the mathematical connection between length, weight, and maximum tension.)

If we combine the results of the last two exercises, we see that the graph of M comes down from infinity as H moves away from zero and that the graph goes back to infinity as H goes to infinity.

HINT 13.12. *Plot the maximal tension M as a function of H using Mathematica for the specific parameter $w = 5$ and the pole at $x = r = 75$.*

Show that each catenary has an optimal sag, that is, a sag and associated horizontal tension H so that the maximum total tension at the pole is minimal amongst all such catenaries.

How can you find $\frac{dM}{dH}$ symbolically in terms of the parameters?

If you want to hang a chain with $w = 5$ in such a way as to minimize the strength required to support it, how much sag do you allow in 75 m?

Chapter 6

Log and Exponential Functions

The temperate zone insect models in the text CD Chapter 20 on Discrete Dynamical Systems use the fact that all the new "bugs" are "born" at the same time. Mayflies hatch in the spring, live a few days, lay eggs and die. The next spring a certain multiple of last spring's population hatches. If each "mother" mayfly has b eggs survive and hatch, then the population next spring will be $p[t+1] = b\,p[t]$, or the change over this spring will be

$$p[t+1] - p[t] = \beta \cdot p[t]$$

with $\beta = b - 1$. If we begin with $p[0] = P_0$, the formula for the solution to the change equation $p[t+1] - p[t] = \beta \cdot p[t]$ is an exponential,

$$p[t] = P_0 \cdot b^t$$

with $b = 1 + \beta$ (until environmemtal limitations must enter the model), but only the values $t = 1, 2, 3, \cdots$ are meaningful in this case.

When every generation is not born simultaneously, generations overlap. In the high school review CD Chapter 28, Problem 28.2 solves the problem of how many algal cells were present in a pond where conditions allowed the cells to divide, and hence double, every six hours. This gives the function

$$N[t] = N_0 2^{t/6}$$

For example, in 6 hours we have $N[6] = N_0 \times 2$, in 12 hours we have twice the population at time 6, $N[12] = N[6] \times 2 = N_0 \times 2 \times 2 = N_0 \times 2^2$, and so forth. In this model it makes perfectly good sense to ask how many algal cells are present after 2 or 3 hours – we do not have to stick to the discrete 6-hour generations, because algal cells in a pond will certainly not all divide simultaneously.

The reason that the formula $N[t] = N_0 \times 2^{t/6}$ also works for other time periods can be understood in the following way. In 3 hours we have

$$N[3] = N_0 \times 2^{3/6} = N_0 \times \sqrt{2}$$

In three more hours we have to multiply our population by the same factor, since the mechanism for growth is the same. This gives

$$N[6] = N[3] \times 2^{3/6} = N[3] \times \sqrt{2} = N_0 \times \sqrt{2} \times \sqrt{2} = N_0 \times 2$$

by the basic law of exponentials $a^x \times a^y = a^{x+y}$. In other words, the result at the end of two half generations of 3 hours each is the doubling we expect after one full six hour generation.

The natural calculus question for our algae model is

$$\text{GIVEN} \quad N[t] = N_0 \times 2^{t/6} \quad \text{FIND the instantaneous rate} \quad \frac{dN}{dt}[t] =?$$

The text CD Section 5.4, text Sections 8.1 and 8.2, and the program **PercentGth** show that the fundamental connection between exponentials and "growth laws" is that the rate of change is proportional to the amount. The limiting case of the identity $a^x \times a^y = a^{x+y}$ is $\frac{dy}{dx} \propto y$. Either way, exponential growth is ubiquitous in science. The projects in this chapter show a few examples.

PROJECT 14

The Canary Resurrected

This project is to try to experimentally verify Newton's law of cooling: the rate of cooling is proportional to the difference between the object temperature and ambient temperature. If we let T = the number of degrees Celsius above ambient and t = the time in minutes, Newton's law of cooling becomes

$$\frac{dT}{dt} = -kT$$

We know that if the initial temperature is T_0, the solution to the differential equation with this initial condition is

$$T = T_0 e^{-kt}$$

There are two questions: (1) How do we measure k? and (2) How good is Newton's law of cooling?

Three students decided to try an experiment and find out for themselves. They put hot water in a plastic cup in a room at 26° C and measured the following temperature differences:

t	T
0	94 − 26 = 68
5	88 − 26
10	84 − 26
15	80 − 26
20	76 − 26
25	73 − 26
30	70 − 26

Scientists often use "semilog" plots of quantities. We want you to discover why.

HINT 14.1. *Preliminary "Semilog" Theory*
Suppose we have a function $T = T_0 e^{-kt}$. Take natural logs of both sides of this equation and show that

$$\tau = \tau_0 - kt$$

where τ is the new variable $\tau = \text{Log}[T]$. The new "semi-log" function is linear. Where does T_0 appear on its graph? How would you measure k from its graph?

HINT 14.2. *Purer Semi-log Linearity*
Suppose we have a function $T = T_0\, e^{-kt}$. Show that
$$\text{Log}\left[\frac{T_0}{T}\right] = k\, t$$
How could you measure k from a plot of $\text{Log}\left[\frac{T_0}{T}\right]$ versus t?

HINT 14.3. *Take logs of T, the temperature differences above ambient for the data above and plot $\text{Log}[T]$ versus t. If Newton's law of cooling is correct, what should the graph look like? How can you measure k on this graph? How would you compute the best approximation to k? How well does the students' data match Newton's law of cooling? Can you think of mistakes they may have made in their experimental procedure?*

Use the program **CoolHelp** from our website if you wish.

You can either just look at this data question or write up a good review problem of the whole sad story of Suzie's canary. Review Problem 4.1, Exercise 4.2.1, Exercise 8.2.1, Problems 8.1, and 8.2 from the main text and write up a complete explanation of Newton's law of cooling for the canary.

PROJECT 15

Drug Concentration and "Biexponential" Functions

Exponential functions arise in the study of the dynamics of drugs in the body. Here is a basic example. Suppose a drug is introduced into the blood stream, say by an intravenous injection. The injection rapidly mixes with the whole blood supply and produces a high concentration of the drug everywhere in the blood. Several tissues will readily absorb the drug when its concentration is higher in the blood than in the tissue, so the drug quickly moves into this second tissue "compartment." The kidneys slowly remove the drug from the blood at a rate proportional to the blood concentration. This causes the blood concentration to drop, and eventually it drops below the tissue concentration. At that point, the drug flows from tissue back into the blood and is continually eliminated from the blood by the kidneys. In the long term, the drug concentration tends to zero in both blood and tissue. The speeds with which these things happen is the subject of "pharmacokinetics."

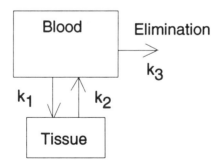

FIGURE 15.1: Two Compartments

Why should we care about such dynamics? Some drugs have undesirable, or even dangerous, side effects if their concentration is too high. At the same time those drugs must be above a certain concentration to be effective for their intended use. As the drug is eliminated from the body, doses need to be given periodically in order to maintain the threshold level for effectiveness, yet doses cannot be too frequent or too large or the concentration will exceed a dangerous level.

For now we just consider a single dose of drug that is introduced into the blood, diffuses into tissue, and is eliminated by the kidneys. Some of the drugs that fit the two compartment model are aspirin (acetylsalicylic acid); creatinine, a metabolite of creatine produced by muscle contraction or degeneration; aldosterone; griseofulvin (an antifungus drug); and lecithin.

The concentrations in blood and tissue respectively are given by a linear combination of two exponential functions (sometimes called "biexponentials" in pharmacokinetics):

$$c_B[t] = b_1 \, e^{-h_1 t} + b_2 \, e^{-h_2 t}$$
$$c_T[t] = -b_3 \, e^{-h_1 t} + b_3 \, e^{-h_2 t}$$

where the positive constants b_1, b_2, b_3, h_1, and h_2 may be computed from physical parameters of the patient.

This project uses these formulas to understand the behavior of such a drug in the body. The Project 35, titled Drug Dynamics and Pharmacokinetics, has you show why the concentrations are given by these exponential formulas. The basic reason is that flow of the drug is described by simple differential equations. Project 16 shows how to use the ideas of this project to measure the dynamic parameters of a patient from data about blood concentrations.

15.1. Primary Variables of the Model

t	– Time, beginning with the initial introduction of the drug, in hours. hr
$c_B = c_B[t]$	– The concentration of the drug in the blood in milligrams per liter. mg/l
$c_T = c_T[t]$	– The concentration of the drug in the tissue in milligrams per liter. mg/l

HINT 15.2. *Show that the units of the drug concentrations in mg/l are equal to units of micrograms per milliliter, so that we might measure the number of micrograms in a milliliter of a patient's blood, rather than the number of milligrams in a liter of blood.*

Each drug and each patient have certain important constants associated with them. In the complete story these will have to be measured.

15.2. Parameters of the Model

v_B	– The volume of the blood compartment in liters. l
v_T	– The volume of the tissue compartment in liters. l
k_1	– A rate constant for diffusion into the tissue compartment.
$k_2 = k_1 v_B / v_T$	– A rate constant for diffusion out of the tissue compartment.
k_3	– A rate constant for elimination by the kidneys.

Concentration is an amount per unit volume. Suppose that a patient has 2.10 liters of blood and that 0.250 grams of a substance is introduced into the patient's blood. When mixed, the concentration becomes $0.25 \times 1000/2.10 = 119$. (mg/l).

HINT 15.3. *Secondary Variables*
Give a general formula to convert the blood and tissue concentrations into the actual total amounts of the drug in the blood or tissue at a given time,

$$a_B = a_T[t] - ??c_B[t] \qquad \text{The amount of the drug in the blood compartment. Units?}$$
$$a_T = a_T[t] = ??c_T[t] \qquad \text{- The amount of the drug in the tissue compartment. Units?}$$

Suppose a patient has $v_B = 2.10$ and $v_T = 1.30$. If $c_B[t]$ and $c_T[t]$ were known functions, what would the amounts $a_B[t]$ and $a_T[t]$ be?

15.3. The Formulas for Concentration

What we know from the theory in Project 35 is that

$$c_B[t] = c_I \left(w_1 \, e^{-h_1 t} + w_2 \, e^{-h_2 t} \right)$$
$$c_T[t] = c_I \left(-w_3 \, e^{-h_1 t} + w_3 \, e^{-h_2 t} \right)$$

where c_I is the initial concentration, or (dose/blood volume) and

$$h_1 = \frac{1}{2}[(k_1 + k_2 + k_3) + \sqrt{(k_1 + k_2 + k_3)^2 - 4\, k_2\, k_3}]$$
$$h_2 = \frac{1}{2}[(k_1 + k_2 + k_3) - \sqrt{(k_1 + k_2 + k_3)^2 - 4\, k_2\, k_3}]$$
$$w_1 = \frac{h_1 - k_2}{h_1 - h_2}, \qquad w_2 = \frac{k_2 - h_2}{h_1 - h_2}, \qquad w_3 = \frac{h_1 - k_2}{h_1 - h_2} \frac{k_2 - h_2}{k_1}$$

These formulas give the exact symbolic unique solution of the full drug dynamics differential equations with initial condition $(c_I, 0)$. They are messy formulas but amount to something that the computer can easily compute for us. We want to get some sort of "feel" for the graph of such functions.

```
k1 = 0.17;
vB = 2.1;
vT = 1.3;
k3 = 0.03;
k2 = k1 vB/vT;
h1 = ((k1 + k2 + k3) + Sqrt[(k1 + k2 + k3)^2 - 4 k2 k3])/2;
h2 = ((k1 + k2 + k3) - Sqrt[(k1 + k2 + k3)^2 - 4 k2 k3])/2;
u1 = vT/(vT + vB)
w1 = (h1 - k2)/(h1 - h2)
u2 = vB/(vT + vB)
w2 = (k2 - h2)/(h1 - h2)
cB[t_] := cI (w1 Exp[ -h1 t] + w2 Exp[ -h2 t]);
cB[t]
```

HINT 15.4. *Graphing Exponentials*
Graph $c_B[t]$ and $c_T[t]$ in case $k_1 = 0.17$, $v_B = 2.10$, $v_T = 1.30$ and $k_3 = 0.03$ and for several other choices of the parameters. What is the general behavior of the graphs? Plot for a period of several days. (Note t is in hours.)

Notice that the blood concentration graph seems to have two parts: A fast decline followed by a slower decline. What physiological things are associated with the fast and slow dynamics in the drug model? Which of the two exponentials decreases fastest?

HINT 15.5. *Semilog Sums of Exponentials*
Graph $\text{Log}[c_B[t]]$ in case $k_1 = 0.17$, $v_B = 2.10$, $v_T = 1.30$, and $k_3 = 0.03$. What is the general behavior of the graphs? Plot for a period of several days. (Note t is in hours.)

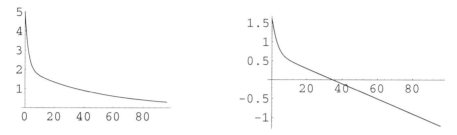

FIGURE 15.6: $c_B[t] = 3\,e^{-0.46\,t} + 2\,e^{-0.02\,t}$ & $\text{Log}[\,c_B[t]\,]$

The first thing we can measure from data is the slow exponential. This comes from the striking feature that you should observe in the semilog plot (or plot of $\text{Log}[c_B[t]]$).

HINT 15.7. *The Log-Linear Tail*
Why is $h_1 > h_2$? (Hint: What is the formula for h_2?)

Given that $h_1 \gg h_2$, if $b_1 \sim b_2$ and $t \gg 0$, how does $b_1\,e^{-h_1 t} + b_2\,e^{-h_2 t}$ compare with $b_2\,e^{-h_2 t}$? You could look at the sizes of both quantities or make the relative measurement and prove

$$\lim_{t \to \infty} \frac{b_2\,e^{-h_2 t}}{b_1\,e^{-h_1 t} + b_2\,e^{-h_2 t}} = \lim_{t \to \infty} \frac{1}{1 + \frac{b_1}{b_2}\,e^{(-h_1 + h_2)\,t}} = 1$$

This limit says that for large t, the term $b_2\,e^{-h_2 t}$ accounts for almost 100% of the concentration. Why does it say this? In any case, explain why the term $b_1\,e^{-h_1 t}$ accounts for the initial fast drop, while the term $b_2\,e^{-h_2 t}$ accounts for the later slow drop and why the later part is mostly $b_2\,e^{-h_2 t}$.

Suppose t is fairly large, so that the fast exponential is negligible, $b_1 e^{-h_1 t} \approx 0$. What is the graph of $\text{Log}[c_B[t]] \approx \text{Log}[b_2 e^{-h_2 t}]$ for an interval of t values in this range? (Hint: Look at the previous figure and justify the graphically obvious feature of the tail of the plot. Use $\text{Log}[x \cdot y] = \text{Log}[x] + \text{Log}[y]$ and $\text{Log}[e^{-h_2 t}] = ??$.)

Show that the slope of the near-linear tail of the $\text{Log}[c_B[t]]$ graph is approximately $-h_2$; in fact, the line has the form

$$\text{Log}[c_B[t]] \approx \beta_2 - h_2 t \quad \text{for a constant } \beta_2, \text{ when } t \gg 0$$

How could you find b_2 by extending the linear tail back to the c_B axis?

Comparison of the model with data will require us to find where the linear tail in the semilog plot begins. A clue is to compare the tissue concentration with the blood concentration:

FIGURE 15.8: Twenty-four Hours of $c_B[t]$, $c_T[t]$, and Log[$c_B[t]$]

The peak in the tissue concentration occurs just about at the end of the fast decline and the start of the log-linear decline. We can find this time.

HINT 15.9. *Maximum Tissue Concentration*
Let $c_T[t] = -w_3\, e^{-h_1 t} + w_3\, e^{-h_2 t}$ for positive constants w_3, h_1, and h_2. Find the maximum of c_T and show that the time where it occurs

$$t_{Max} = \frac{\mathrm{Log}[h_1] - \mathrm{Log}[h_2]}{h_1 - h_2}$$

Compute this time for the specific constants w_3, etc. coming from the model parameters k_1, etc. where you have already graphed $c_T[t]$. Compare this time with the peak on your graph.

The overall effect of a drug is related to the integral of $c_B[t]$ during the time when it remains above a minimum concentration for effectiveness, c_E.

HINT 15.10. *Total Effectiveness of a Single Dose*
Show that

$$\int_0^{t_E} c_B[t]\, dt = c_I \left(\frac{w_1}{h_1}\left(1 - e^{-h_1 t_E}\right) + \frac{w_2}{h_2}\left(1 - e^{-h_2 t_E}\right) \right)$$

and

$$\int_{t_1}^{t_2} c_T[t]\, dt = c_I \left(\frac{w_3}{h_1}\left(e^{-h_1 t_2} - e^{-h_1 t_1}\right) + \frac{w_3}{h_2}\left(e^{-h_2 t_1} - e^{-h_2 t_2}\right) \right)$$

You may do your computation with the computer or the help of the program **BiExponentialHelp** on our website.

HINT 15.11. *Finding Threshold Times and Cumulative Effects*
Use numerical values of w_1, etc. from the exact solution of the model for choices of the parameters to find the approximate time t_E where $c_B[t_E] = \frac{1}{4} c_I = c_E$. For example, use Mathematica's **FindRoot[cB[t] == cE , {t,3}]** *or the program* **BiExponentialHelp** *on our website.*

Then compute the numerical value of $\int_0^{t_E} c_B[t]\, dt$. What are the units of the integral?
Find two times t_1 and t_2 where $c_T[t] = \frac{1}{4} cI$. Why are there two such times? Then compute the numerical value of $\int_{t_1}^{t_2} c_T[t]\, dt$.

15.4. Comparison with Mythical Data

The mythical drug mathdorphin (MD) is produced in the body in large doses after a long period of serious effort on a difficult but interesting task. When the concentration of MD in the brain exceeds 1 mg/l it produces a state of elated satisfaction. Furthermore, the patient's IQ is noticeably increased in proportion to the period of time during which this excess concentration is maintained.

The amount of MD produced by the body is proportional to the square of the effort times the difficulty of the task. Mathdorphin is released into the blood stream when a first submission of a project is handed in. In this case the blood compartment consists of blood, liver, lungs, kidney tissue, endocrine glands, muscle, adipose, marrow, and skin. A typical student then has a "blood" volume of 13.4 l. The "tissue" compartment consists of the brain and spinal compartment, where the important action of MD occurs. The typical "tissue" compartment in this case is 2.31 l.

Through careful observation of students, we have collected rate parameters for a typical calculus student: $k_1 = 0.25$ and $k_3 = 0.075$.

EXPLORATION 15.12. *Cumulative and Peak Effects of MD*
In the case of MD, the concentration in the brain must be held above 1 mg/l to produce the beneficial effect. For each mg/l above 1 maintained 36 hours, the patient's IQ increases by one point. How much smarter is this student after working this project?

MD does have one dangerous side effect. If the concentration is maintained above 10 mg/l for more than 5 hours, the patient develops an irresistible urge to attend graduate school in mathematics. Will you develop this neurosis? Try various initial blood concentrations and see how high this can go before you become a math nerd. How much is your IQ increased?

15.5. Comparison with Real Data

The next project uses the ideas of this project to show how the rate parameters may be measured from real data.

PROJECT 16

Measurement of Kidney Function by Drug Concentration

This project is the empirical part of our three projects on drug dynamics, or "pharmacokinetics." The primary quantities we wish to study are the drug concentration in the blood and in the tissue. These concentrations vary as a functions of time. Project 35 shows theoretically why a sum of two exponential functions gives the blood and tissue concentrations of a drug, while Project 15 studies some of the basic mathematics of these "biexponential" functions. This project ties the real measured concentrations with the theoretical model. We use Hint 15.7 to measure the parameters of a real patient's response to a drug.

16.1. Variables and Parameters

We will use the same variables and parameters in this project as in Project 15 and Project 35. You should read the introductions to those projects for background ideas, but not worry about the mathematical details of those projects. They show the following:

PROJECT 35 SHOWS: that for drugs that satisfy the "linear two compartment model," the concentrations in blood and tissue respectively after a single intravenous injection are modeled by a linear combination of two exponential functions:

$$c_B[t] = b_1\ e^{-h_1 t} + b_2\ e^{-h_2 t}$$
$$c_T[t] = -b_3\ e^{-h_1 t} + b_3\ e^{-h_2 t}$$

The positive constants b_1, b_2, b_3, h_1, and h_2 may be computed from physical parameters of the patient. (Formulas for b_1, etc. are given in Projects 15, 35, and below in this project.)

PROJECT 15 SHOWS: the graphs and behavior of such biexponential functions. In particular, Hint 15.7 shows that the semilog graph of $c_B[t]$ (or graph of Log[$c_B[t]$]) has a linear tail. This is the starting point for our empirical work.

16.2. Overview of the Project

The theoretical blood concentration has the analytical form

$$c_B[t] = b_1\ e^{-h_1 t} + b_2\ e^{-h_2 t}$$

The rate constant h_2 is related to kidney function and the rate constant h_1 to absorption by tissue. Typically, $h_2 \ll h_1$.

An ideal semilog plot of blood concentration data after a single injection looks like Figure 16.1.

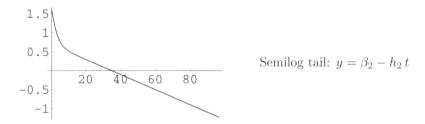

FIGURE 16.1: Log of Blood Concentration versus Time

Notice the rapid drop followed by the (log-)linear decline. According to Hint 15.7, the tail of this graph is the linear function $y = \beta_2 - h_2 t$ where $\beta_2 = \text{Log}[b_2]$. In other words, the slope $-h_2$ is the rate constant associated with kidney function and the y-intercept is the log of the constant b_2. This gives a way to measure b_2 and h_2 from data. Once measured, if we subtract this measured part from the data, the resulting semilog plot is another line of slope $-h_1$ with intercept $\text{Log}[b_1]$, corresponding to the fast initial drop.

16.3. Drug Data

Figure 16.2 is a graph of blood concentration data similar to human data for the anticoagulant warfarin. (We could not obtain the raw data for a real human, so approximated them.) Our actual numbers are contained in the program **DrugData** on our website.

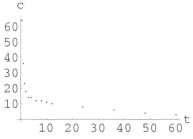

FIGURE 16.2: Warafarin Data

The data are given in a list of time-concentration pairs (t, c). These are entered in the form

data = {{.5,64.},{1,36.},{1.5,23.},{2,18.}, {3,14.},{4,14.},{6,12.},
    {8,12.},{10,11.},{12,10.},{24, 8.},{36,6.},{48,4.0},{60,3.0}};

The times and concentrations can be separated as lists in *Mathematica* as follows:

In[2]
  {times, concens} = Transpose[data];
  times
  concens
Out[2]

{0.5, 1, 1.5, 2, 3, 4, 6, 8, 10, 12, 24, 36, 48, 60}
{64., 36., 23., 18., 14., 14., 12., 12., 11., 10., 8., 6., 4., 3.}

Once concentrations are separated, we can make a semilog plot as follows:
logCons = Log[concens];
logData = Transpose[times,logCons];
logPlot = ListPlot[logData,AxesLabel-¿"t","Log[c]"];

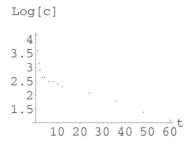

FIGURE 16.3: Log of Concentration versus time

Now we have the problem of guessing which points constitute the log-linear tail. We can take the last eight data points as follows:
tail8 = Take[logData, -8]
These can be fit to a straight line by entering:
f8 = Fit[tail8,{1,t},t]
tail8Plot = Plot[f8,{t,0,60},AxesLabel->{"t","Log[c]"}];
Show[tail8Plot,logPlot];
Fits for the last 8, last 6 and last 4 points are shown next:

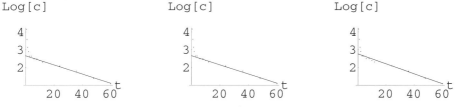

FIGURE 16.4: Linear Fits for 8, 6 and 4 Points

EXPLORATION 16.5. *How many points should you use to fit the data to the linear tail? What are the various values of β_2 and h_2 in the fits,*

$$f = \beta_2 - h_2 t$$

Once you have chosen β_2 and h_2 from your best fit, you need to go back to the original data and remove the tail portion. This can be done on the whole list with *Mathematica* as follows:
In[19]
  h2 = 0.026
  b2 = Exp[2.7]
  remainders = concens - b2 Exp[- h2 times]
Out[19]
  {49.3125, 21.5022, 8.68941, 3.87424, 0.236777, 0.590009,
  -0.730491, -0.0854228, -0.473041, -0.891689, 0.027485, 0.164266, -0.271649, -0.126768}

EXPLORATION 16.6. *What is wrong with some of the remainders shown above? Can we take logs of the bad ones? What might have caused the trouble?*

Our concentrations ideally are of the form
$$c_B[t] = b_1 \, \text{Exp}[-h_1 \, t] + b_2 \, \text{Exp}[-h_2 \, t]$$
for the times in out list. You have measured b_2 and h_2 the best you could from the data and computed the list of remainders.

EXPLORATION 16.7. *As a function of t, what is the symbolic form of the remainders list?*
$$c_R[t] = c_B[t] - b_2 \, \text{Exp}[-h_2 \, t] = ??$$
Show that the symbolic form of the log of the remainders expression is a linear function
$$\text{Log}[c_R[t]] = \beta_1 - h_1 \, t$$
What are β_1 and h_1 in terms of the parameters of the problem?

Ignoring the bad data for the moment, we can form the logs of the remainder data as follows in *Mathematica*:

headData = Transpose[{times,Log[remainders]}]

We can plot the first 6 points as follows:

headPlot = ListPlot[Take[headData,6],
  AxesOrigin->{0,0},AxesLabel->{"t","Log[c]"}];

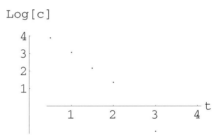

FIGURE 16.8: Six Points of $\text{Log}[c_B[t] - b_2 \, e^{-h_2 t}]$

We can fit a line to the first four points as follows:

g = Fit[Take[headData,4],1,t,t]
headLine = Plot[g,t,0,4,AxesLabel->{"t","Log[c]"}];
Show[headLine,headPlot];

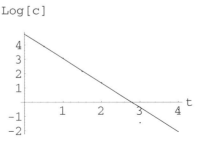

FIGURE 16.9: Linear Fit of the First 4 Remainder Points

EXPLORATION 16.10. *The linear fit tells us b_1 and h_1. How?
Which is the best fit to the initial drop in concentration, the fit to 4, 5, 6 or 7 points?*

Once you have answered this question, you know the approximate blood concentration function; for example, you might type:
b1 = Exp[4.8];
h1 = 1.7;
cA[t_] := b1 Exp[-h1 t] + b2 Exp[-h2 t];
cA[t]
curvePlot = Plot[cA[t],{t,0,60}];
Show[curvePlot,dataPlot];

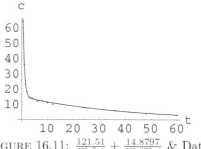

FIGURE 16.11: $\frac{121.51}{E^{1.7\,t}} + \frac{14.8797}{E^{0.026\,t}}$ & Data

We could stop the project here, since we have measured the constants in the symbolic solution of the blood concentration. However, it would be nice to connect our measured values of b_1, b_2, h_1, and h_2 with the physical parameters k_1, v_B, v_T, k_3, and c_I, the initial concentration. (See Project 15 or Project 35 for the definitions of these parameters.)

HINT 16.12. *The initial dose of warfarin that produced the patient response of our data was the amount $a_I = 100$ mg. What is $c_B[0]$ with your measured values of b_1, etc.? What is the ratio $c_I = a_I/v_B = 100/v_B$? What is v_B?*

Two down, three to go. Remember that $k_2 = k_1 v_B/v_T$.

HINT 16.13. *The Physical Parameters versus The Measured Constants*
We are given that the ideal blood concentration is
$$c_B[t] = b_1\,e^{-h_1\,t} + b_2\,e^{-h_2\,t}$$
where
$$h_1 = \frac{1}{2}[(k_1 + k_2 + k_3) + \sqrt{(k_1 + k_2 + k_3)^2 - 4\,k_2\,k_3}]$$
and
$$h_2 = \frac{1}{2}[(k_1 + k_2 + k_3) - \sqrt{(k_1 + k_2 + k_3)^2 - 4\,k_2\,k_3}]$$

1) Show that
$$h_1\,h_2 = k_2\,k_3 \quad \text{and} \quad h_1 + h_2 = k_1 + k_2 + k_3$$

2) We are also given that $b_1 = c_I \, w_1$ and $b_2 = c_I \, w_2$, where

$$w_1 = \frac{h_1 - k_2}{h_1 - h_2} \qquad \& \qquad w_2 = \frac{k_2 - h_2}{h_1 - h_2}$$

We already know the values of b_1, b_2 and c_I, so give the values of $w_1 = ?$ and $w_2 = ?$.

3) We also know measured values of h_1 and h_2, so compute k_2 with both of the formulas above and find v_T.

4) Above you showed that $h_1 \, h_2 = k_2 \, k_3$, and now you know h_1, h_2, k_2. Compute $k_3 = ?$.

5) Above you showed that $h_1 + h_2 = k_1 + k_2 + k_3$, and now you know h_1, h_2, k_2, and k_3. Compute $k_1 = ?$.

6) Compute $v_T = ?$.

PROJECT 17

Numerical Derivatives of Exponentials

We start with the question

$$\text{GIVEN} \quad y[t] = b^t \quad \text{FIND} \quad \frac{dy}{dt}[t] = ?$$

by direct computation. The increment approximation, or microscope equation

$$f[x + \delta x] = f[x] + f'[x] \cdot \delta x + \varepsilon \cdot \delta x$$

becomes

$$y[t + \delta t] = y[t] + y'[t] \cdot \delta t + \varepsilon \cdot \delta t$$

specifically

$$b^{t+\delta t} = b^t + y'[t] \cdot \delta t + \varepsilon \cdot \delta t$$

with $y'[t]$ and ε unknown. We solve for the unknowns, obtaining

$$y'[t] + \varepsilon = \frac{b^{t+\delta t} - b^t}{\delta t} = b^t \frac{b^{\delta t} - 1}{\delta t}$$

(using the exponential identity $b^{t+\delta t} = b^t \cdot b^{\delta t}$). Assuming that $\varepsilon \approx 0$, we have

$$y'[t] = \frac{db^t}{dt} \approx b^t \cdot \frac{b^{\delta t} - 1}{\delta t} = y[t] \cdot \frac{b^{\delta t} - 1}{\delta t}$$

or

$$\frac{1}{y[t]} \cdot \frac{dy}{dt}[t] \approx \frac{b^{\delta t} - 1}{\delta t} \quad \text{when} \quad \delta t \approx 0$$

We want to estimate the real constant k_b so that

$$\frac{b^{\delta t} - 1}{\delta t} \approx k_b \quad \text{when} \quad \delta t \approx 0, \quad \text{or equivalently,} \quad \lim_{\Delta t \to 0} \frac{b^{\Delta t} - 1}{\Delta t} = k_b$$

Then we have the derivative formula.

$$\frac{1}{b^t} \cdot \frac{db^t}{dt} = k_b \quad \Leftrightarrow \quad \frac{1}{y[t]} \cdot \frac{dy}{dt}[t] = k_b$$

Begin your project with a derivation of this formula.

HINT 17.1. *Show that*

$$\text{If } y[t] = b^t, \text{ then } \frac{dy}{dt}[t] = k_b\, y[t] \quad \text{or} \quad \frac{db^t}{dt} = k_b\, b^t$$

where

$$\frac{b^{\delta t} - 1}{\delta t} \approx k_b \quad \text{with} \quad \delta t \approx 0, \quad \text{or} \quad \lim_{\Delta t \to 0} \frac{b^{\Delta t} - 1}{\Delta t} = k_b$$

Next, use the computer to find a practical approximation to the constant k_2. (Of course, we would also like to know the exact value of the constant, which turns out to be the natural base log of 2, $k_2 = \text{Log}[2]$. The main text, Section 6.5, shows you the exact symbolic way to differentiate exponentials.)

HINT 17.2. *Compute a table of values of*

$$\frac{2^{\Delta t} - 1}{\Delta t}$$

for a sequence of smaller and smaller values of Δt; *for example, you might let* $\Delta t = 1/2.0$, $1/4.0$, $1/8.0$, ... *or* $\Delta t = 0.1, 0.01, 0.001, \ldots$. *Notice that you cannot let* Δt *get "too" small and maintain numerical accuracy in a floating point computation.*

The program **ExpDeriv** on our website has help with these computations.

Compare your approximation with the computer's high-precision value of e. Your computation can only claim a reasonable amount of accuracy. How much is reasonable?

You can improve the accuracy by modifying the program to compute symmetric differences

$$\frac{y[t + \delta t/2] - y[t - \delta t/2]}{\delta t} \approx y'[t]$$

rather than $[y(t+\delta t) - y[t]]/\delta t$. Experimentally, you will see that this is more accurate. The Mathematical Background chapter on Taylor's formula, Section 8.2, gives the mathematical reason for this accuracy.

HINT 17.3. *Write the increment approximation for the function* $y[t]$ *with both* δt *and* $-\delta t$,

$$y[t + \delta t] = y[t] + y'[t] \cdot \delta t + \varepsilon_1 \cdot \delta t$$
$$y[t - \delta t] = y[t] - y'[t] \cdot \delta t + \varepsilon_2 \cdot \delta t$$

and then subtract the equations and solve for $y'[t]$. *Use this formula on the specific function* $y[t] = b^t$ *to show that*

$$\frac{1}{y[t]} \cdot \frac{dy}{dt} \approx \frac{b^{\delta t} - 1/b^{\delta t}}{2\,\delta t}$$

Use this formula to compute a table of closer and closer approximations to k_2 *and compare the accuracy with your first computation.*

Once you have confidence in the accuracy of your tables of approximations for k_2, compute tables of k_b for the following exponential bases, b.

HINT 17.4. *Use the* **ExpDeriv** *computer program to show that* $k_2 \approx .693$, $k_3 \approx 1.099$, $k_{2.7} \approx .993$, $k_{2.8} \approx 1.03$. *Can you claim more accurate values for these constants with the computations in your program?*

HINT 17.5. *Euler's Number e as the Base with Constant $k_e = 1$*
Use the computer to find the number b that has $k_b = 1$. This is the "natural" base to use in calculus.

We haven't really proved that the limits defining the constants k_b exist, but the numerical data from the the computer program are pretty convincing evidence that they do. The meaning is quite important, namely we see that

$$\text{If } y[t] = b^t, \text{ then } \frac{dy}{dt}[t] = k_b\, y[t] \text{ or } \frac{db^t}{dt} = k_b\, b^t$$

Moreover, you should see that there is some constant e that makes the proportionality constant $k_e = 1$. This is Euler's first constant $e \approx 2.71828$ and it therefore satisfies the important identity

$$\text{If } y[t] = e^t, \text{ then } \frac{dy}{dt}[t] = y[t] \quad \text{or} \quad \frac{de^t}{dt} = e^t$$

We use the relation $\frac{dy}{dt} = y$ as a cornerstone of our "official" theory of exponentiation in the text. The fact that the constant of proportionality is one for the base e is why e is considered the "natural" base for logs and exponentials. This is somewhat like radian measure for angles, where at first you may prefer degrees but only get $\frac{d\,\mathrm{Sin}[\theta]}{d\theta} = \mathrm{Cos}[\theta]$ in radians.

HINT 17.6. *Unnatural Angles*
The sine function in degrees is given by $S[D] = \mathrm{Sin}[\pi D/180]$. Explain. Show that the derivative of the sine function in degrees is approximately 0.01745 times the cosine function in degrees.

PROJECT 18

Repeated Exponents

The number $b^{(b^b)}$ need not equal $(b^b)^b$. For example, $3^3 = 3 \cdot 3 \cdot 3 = 27$, so $3^{(3^3)} = 3^{27} = 7625597484987$, but $(3^3)^3 = 3^{(3\cdot 3)} = 3^9 = 19683$, or $(3^3)^3 = 27^3 = 19683$. "Exponentiation is not associative."

The limit

$$b, \quad b^b, \quad b^{b^b}, \quad b^{b^{b^b}}, \quad b^{b^{b^{b^b}}} \cdots$$

as the number of exponentiations tends to infinity is an interesting one. This can be viewed as a discrete dynamical system, (as in CD text Chapter 20).

HINT 18.1. *Let $g[x] = b^x$ and define a sequence $a_0 = 1$, $a_1 = g[a_0]$, $\cdots$, $a_{n+1} = g[a_n]$. Compute a_1, a_2, a_3, a_4, and a_5 explicitly.*

In function notation, the sequence is repeated computation of the function,
$a_0 = 1$, $a_1 = g[a_0] = g[1]$, $a_2 = g[a_1] = gg[g[1]]$, $a_3 = g[a_2] = g[g[g[1]]]$, $\cdots$

In CD Chapter 20 of the main text, these "discrete dynamical systems" are studied in some detail. The program **FirstDynSys** from CD Chapter 20 can be used to compute and graph the sequence. Try it for $b = 1.2$, $b = 0.8$ and some other values. (Make the function $g[p] = b^p$ and $f[p] = g[p] - p$. Then $a_1 = b^1$, $a_2 = b^b$, $a_3 = b^{a_2} = b^{b^b}$, ...)

If the limit

$$b, \quad b^b, \quad b^{b^b}, \quad b^{b^{b^b}}, \quad b^{b^{b^{b^b}}} \cdots$$

equals a_∞, then we must have $g[g[\cdots g[1]\cdots]] \approx a_\infty$ and $g[g[g[\cdots g[1]\cdots]]] \approx g[a_\infty] \approx a_\infty$.

HINT 18.2. *Show that the equation $g[x] = x \Leftrightarrow x = b^x$ has no solutions if $b = 3$. Start with a plot,*

Plot[{3∧x, x},{ x, 0, 2}]

HINT 18.3. *Show graphically that the equation $g[x] = x \Leftrightarrow x = b^x$ has solutions if $b = 1.1$ or if $b < 1$.*

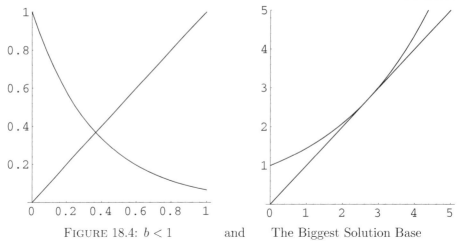

FIGURE 18.4: $b < 1$ and   The Biggest Solution Base

We would like to know the value of b where $y = b^x$ just touches $y = x$. Larger values of b give no solutions to $x = b^x$.

HINT 18.5. *The Biggest b*
Show that $\dfrac{\text{Log}[x]}{x} = \text{Log}[b]$ *is equivalent to the equation* $x = b^x$.

Use the computer to graph $\dfrac{\text{Log}[x]}{x}$.

Prove that the maximum of $\dfrac{\text{Log}[x]}{x}$ *is* $1/e$.

Use what you have so far to prove that $b = e^{(1/e)} \approx 1.44467$ is the largest b so that $x = b^x$ has a solution.
Run **FirstDynSys** with $b = e^{(1/e)}$.

Theorem CD 20.4 of the main text gives a condition for the sequence $g[g[g[\cdots]]]$ to converge (when we start close enough to the limit.) It simply says we must have $|g'[x_e]| < 1$ where $g[x_e] = x_e$. Now we want to know the value of the base $b < 1$ that crosses $y = x$ at slope -1.

HINT 18.6. *The Lower Limit*
Calculate the derivative $g'[x]$ and show that Theorem CD 20.4 says we must have

$$x_e = b^{x_e} \quad \text{and} \quad |b^{x_e} \, \text{Log}[b]| < 1$$

Solve the pair of equations $x_e = b^{x_e}$ and $|b^{x_e} \, \text{Log}[b]| = 1$ with the hints:
The first equation gives $\text{Log}[x_e] = x_e \, \text{Log}[b]$.
This means $x_e = 1/e$. *Why?*
Substitute $x_e = 1/e$ *into the first equation to show that* $b = 1/e^e$.
Calculate $1/(e^e) \approx 0.06 \cdots$ *using the computer.*
Run **FirstDynSys** with $b = 0.065988$ for 1000 terms of the sequence.
Run **FirstDynSys** with $b = 0.06$ for 1000 terms of the sequence.

There is much more on this problem in the American Math Monthly, vol. 88, nr. 4, 1981, in the article, Exponentials Reiterated, by R. Arthur Knoebel.

PROJECT 19

Solve $dx = r[t]\, x[t]\, dt + f[t]$

The differential equation $dx = r\, x[t]\, dt$ has solutions of the form $x[t] = x_0\, e^{rt}$ when r is constant. (See the main text, Section 8.2.) This project shows that you can still solve a differential equation that is "linear in x,"

$$dx = r[t] x[t] dt$$

One "method" is by guessing that the solution has the form

$$x[t] = x_0\, e^{R[t]}$$

for some unknown function $R[t]$.

HINT 19.1. *Let $x[t] = x_0\, e^{R[t]}$, for a constant x_0 and unknown function $R[t]$. Use the Chain Rule to show that*

$$\frac{dx}{dt} = x_0\, r[t]\, e^{R[t]} = r[t]\, x[t]$$

provided $R'[t] = r[t]$ or $R[t] = \int r[t]\, dt$.

In particular, show that when $R[t] = \frac{t^2}{2}$, $x[t] = e^{t^2/2}$ sartisfies

$$\frac{dx}{dt} = t\, x[t]$$

Another method of solution is called "separation of variables" and is studied in Chapter 21 of the main text. Here is how that works in this case:

HINT 19.2. *Write the differential equation*

$$\frac{dx}{dt} = \text{Sin}[t] x[t] \quad \text{in the form} \quad \frac{dx}{x} = \text{Sin}[t]\, dt$$

and antidifferentiate both sides with respect to the separate variables,

$$\int \frac{1}{x}\, dx = \int \text{Sin}[t]\, dt$$

$$\text{Log}[x] = -\text{Cos}[t] + k, \quad \text{for any constant } k$$

Solve for x and show that you get the solution

$$x[t] = x_0 \, e^{R[t]} \quad \text{with } x_0 = e^k \text{ and } R[t] = \int \text{Sin}[t] \, dt$$

(Hint: Why does $e^{-\text{Cos}[t]+k} = e^{-\text{Cos}[t]} \cdot e^k$? If k is an arbitrary constant, so is e^k – except for sign.)

Verify by substitution that $x[t] = x_0 \, e^{-\text{Cos}[t]}$ *satisfies the differential equation for any value of the constant* x_0.

Differential equations say how a quantity changes with time. In order to know "where we end up," we need to know where we start.

HINT 19.3. *Solve the initial value problem*

$$x[0] = 3$$
$$\frac{dx}{dt} = \frac{x[t]}{(\text{Cos}[t])^2}$$

(Hint: The derivative of $\text{Tan}[t]$ is $1/(\text{Cos}[t])^2$. What is $x[0]$ if $x[t] = x_0 \, e^{\text{Tan}[t]}$?)

Once we know how to solve $dx = r[t] \, dt$, we can solve the "forced" equation by the unlikely method called "varying the constant." Here's how that works for the "forcing" function $f[t]$.

HINT 19.4. *Suppose we are given the functions $R[t]$ and $r[t]$ where $x[t] = e^{R[t]}$ is a solution of $dx = r[t] \, x[t] \, dt$, that is, $R'[t] = r[t]$. We want to solve the equation*

$$\frac{dx}{dt} - r[t] \, x[t] = f[t]$$

"Guess" the solution $x[t] = c[t] \, e^{R[t]}$ and substitute it into the equation. Use the Product Rule and Chain Rule to show that

$$x'[t] = c'[t] \, e^{R[t]} + r[t] \, x[t] \quad \text{or} \quad \frac{dx}{dt} - r[t] \, x[t] = c'[t] \, e^{R[t]}$$

This "guess" will work provided

$$c'[t] = f[t] \, e^{-R[t]} \quad \text{or} \quad c[t] = \int f[t] \, e^{-R[t]} \, dt$$

HINT 19.5. *Test this "guess" on the equation $\frac{dx}{dt} = t \, x + t$ where $r[t] = t$ and $R[t] = t^2/2$. Compute*

$$c[t] = \int t \, e^{-t^2/2} \, dt = k - e^{-t^2/2}, \quad \text{for constant } k$$

Show that this makes

$$x[t] = k \, e^{t^2/2} - 1$$

and substitute it in the differential equation to verify that it is correct.

The exact differential equation solvers in Maple and *Mathematica* will also solve these equations.

Chapter 7

Theory of Derivatives

PROJECT 20

The Mean Value Math Police

The following situation illustrates the main result of this chapter. You travel a total distance of 111 miles between toll booths in an elapsed time of one and one half hours. All the toll booths have synchronized clocks accurate to within a few seconds. When you arrive at the exit booth, the officer hands you a speeding ticket. He cites you for 74 mph even though, while you were in his sight you were obeying the 65 mph limit and even slowed cautiously before the toll booth. Were you really speeding? Of course, but how can he prove this?

The officer computes your overall (mean or average) speed as $\frac{111 \text{mi}}{3/2 \text{hr}} = 74$ mph and feels he can prove you had to be going that fast at at least two times. You actually had to drive over 74 mph part of the time, didn't you? Do you pay the ticket or will you try to argue in court that you averaged 74 mph without ever going that fast?

20.1. The Mean Value Theorem for Regular Derivatives

We want to formulate this speed problem in a general way for a function $y = f[x]$ on an interval $[a, b]$. You may think of x as the time variable with $x = a$ at the start of the trip and $x = b$ at the end. The elapsed time traveled is $b - a$, or 3/2 hours in the example. (Perhaps you start at 2 and end at 3:30.) You may think of $y = f[x]$ as a distance from a reference point: your odometer reading. We start at $f[a]$, end at $f[b]$, and travel a total of $f[b] - f[a]$. The elapsed time is $b - a$, so the overall "average" speed is $(f[b] - f[a])/(b - a)$. Instantaneous speed is the time rate of change of distance or derivative $f'[x]$.

HINT 20.1. *Sketch the graph of a trip beginning at 2 pm, 35 miles from a reference point, ending at 3:30 pm 146 miles from the reference point and having the features of starting from and stopping at toll booths, as described above. Perhaps you even stop at a rest area.*

Sketch the line connecting the end points of the graph $(a, f[a])$ and $(b, f[b])$. What is the slope of this line?

Find a point on your sketch where the speed is 74 mph and sketch the tangent line at that point. Call the point c. Why does this satisfy $f'[c] = \frac{f[b] - f[a]}{b-a}$?

The following is what the officer needs in court, since he needs to show that there is a time c when your speed $f'[c] = 74$. We already know that $(f[b] - f[a]) = 111$ and $(b - a) = 3/2$, so $(f[b] - f[a])/(b-a) = 74$.

THEOREM 20.2. *The Mean Value for Traffic Court*
Let $f[x]$ be a function that is differentiable at every point of the closed interval $[a,b]$. There is a point c in the open interval $a < c < b$ such that

$$f'[c] = \frac{f[b] - f[a]}{b - a}$$

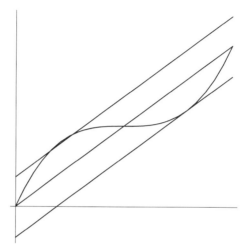

FIGURE 20.3: Mean Slope and Tangents

There may be more than one point where $f'[c]$ equals the mean speed or slope.

PROOF

The average speed over a subinterval of length Δx is

$$g[x] = g_{\Delta x}[x] = \frac{f[x + \Delta x] - f[x]}{\Delta x}$$

and this new function is defined and continuous on $[a, b - \Delta x]$.

Suppose we let $\Delta x_1 = (b-a)/3$ compute the average of 3 averages, the speeds on $[a, a + \Delta x_1]$, $[a + \Delta x_1, a + 2\Delta x_1]$, and $[a + 2\Delta x_1, a + 3\Delta x_1]$. This ought to be the same as the overall average and the telescoping sum below shows that it is:

$$\frac{1}{3}[g[a] + g[a + \Delta x_1] + g[a + 2\Delta x_1]] =$$
$$\frac{1}{3}\left[\frac{f[a + \Delta x_1] - f[a]}{\Delta x_1} + \frac{f[a + 2\Delta x_1] - f[a + \Delta x_1]}{\Delta x_1} + \frac{f[a + 3\Delta x_1] - f[a + 2\Delta x_1]}{\Delta x_1}\right]$$
$$= \frac{f[a + \Delta x_1] - f[a] + f[a + 2\Delta x_1] - f[a + \Delta x_1] + f[a + 3\Delta x_1] - f[a + 2\Delta x_1]}{3\Delta x_1}$$
$$= \frac{-f[a] + f[b]}{3\frac{b-a}{3}} = \frac{f[b] - f[a]}{b - a}$$

This implies that there is an adjacent pair of subintervals with

$$g[x_{lo}] = \frac{f[x_{lo} + \Delta x_1] - f[x_{lo}]}{\Delta x_1} \leq \frac{f[b] - f[a]}{b - a} \leq \frac{f[x_{hi} + \Delta x_1] - f[x_{hi}]}{\Delta x_1} = g[x_{hi}]$$

because the average of the three subinterval speeds equals the overall average and so either all three also equal the overall average, or one is below and another is above the mean slope. (We know that x_{lo} and x_{hi} differ by Δx_1, but we do not care in which order they occur $x_{lo} < x_{hi}$ or $x_{hi} < x_{lo}$.)

The officer wants to preempt an argument on your part that you were not speeding near the toll booths, so he does not want to argue that $g[x_{hi}]$ is either the first or last half-hour. He also has only cited you for 74 mph, so he would like to argue that there was a half-hour interval inside your trip interval where you averaged exactly 74 mph.

He knows
$$g[x_{lo}] \leq \frac{f[b] - f[a]}{b - a} \leq g[x_{hi}]$$
and wants to reason that $g[x]$ attains the intermediate value (of the mean speed, 74) for some x between x_{lo} and x_{hi}. (If your middle half-hour speed is 74, we could use that one, but the officer knows you may have stopped at the Midway Rest Area and have witnesses.)

We will come back to your speed once we establish the fact that there is an x_1 where $g[x_1] = \frac{f[b]-f[a]}{b-a}$.

20.2. The Theorem of Bolzano

Your mean speed $g[x]$ was below 74 at time x_{lo} and above 74 at time x_{hi}, one half hour later. Was $g[x_0]$ ever equal to 74? Most people would say, "Yes." They implicitly reason that speed "moves continuously" through values and hence hits all intermediate values. This idea is a precise mathematical theorem, and its most difficult part is in correctly formulating what we mean by "continuous" function. Our definition was given in Chapters 4 and 5 of the main text. In court, the officer cites the following result to prove that there was a half-hour interval – away from the toll booths, between x_{lo} and x_{hi} – when your average speed was exactly 74. We'll come back to the officer after we see why he's gotcha on this one.

THEOREM 20.4. *Bolzano's Intermediate Value Theorem*
If $y = g[x]$ *is continuous on the interval* $\alpha \leq x \leq \beta$, *then* $g[x]$ *attains every value intermediate between the values* $g[\alpha]$ *and* $g[\beta]$. *In particular, if* $g[\alpha] < 74$ *and* $g[\beta] > 74$, *then there is an* x_0, $\alpha < x_0 < \beta$, *such that* $g[x_0] = 74$.

A complete proof is in the Mathematical Background Section 4.3. The related Theorem of Darboux for derivatives is in the Math Background Section 7.2.

HINT 20.5. *Show that the function* $y = j[x] = \frac{\sqrt{x^2+2x+1}}{x+1}$ *equals* -1 *when* $x = -2$, *equals* $+1$ *when* $x = +3$, *but never takes the value* $y = \frac{1}{2}$ *for any value of* x. *Why doesn't* $j[x]$ *violate Bolzano's Theorem?*

Since $g[x]$ is continuous, Bolzano's Intermediate Value Theorem says that there is an x_1 between x_{lo} and x_{hi} with $g[x_1] = (f[x_1 + \Delta x_1] - f[x_1])/\Delta x_1 = (f[b] - f[a])/(b - a)$ ($= 74$ in your case). The subinterval $[x_1, x_1 + \Delta x_1]$ lies inside (a, b), has length $(b - a)/3$ and $f[x]$ has the same mean slope over the subinterval as over the whole interval. (So far we have only used continuity of $f[x]$.)

You see your opening argument forming. All the officer has done is to show that there is a half-hour interval with the average speed of 74. You will introduce your witness from the Midway Rest Area to show you weren't speeding when you stopped there and argue on the subinterval that you somehow averaged 74 without ever going that instantaneous speed, just the agrument you were planning for the whole trip, but now modified to the interval one third as long.

But the officer has had calculus and his argument isn't finished. He has an interval $[x_1, x_1 + \Delta x_1]$ with $\Delta x_1 = (b-a)/3$ where

$$\frac{f[x_1 + \Delta x_1] - f[x_1]}{\Delta x_1} = \frac{f[b] - f[a]}{b - a} \quad (= 74 \text{ in your case})$$

He has two ways to proceed. One way would be to break this interval into three more and apply the same reasoning to get an interval of one ninth the original time period where your speed was again 74. Another is to break this interval into tiny subintervals and average the speeds over the tiny subintervals.

The officer continues, "Your honor, I will now show that the defendant's speed was 74 mph to within the accuracy of any known measuring device." He is about to show that there is a time x_m and a tiny change in time δx so that

$$\frac{f[x_m + \delta x] - f[x_m]}{\delta x} \approx 74$$

Why does this show that $f'[c] = 74$ for some c?

HINT 20.6. *One Third Approach*
How does the successive intervals argument of length one third the previous interval lead to the conclusion that there is a c with $f'[c] = \frac{f[b]-f[a]}{b-a}$?

HINT 20.7. *Average of Many Tiny Subintervals*
How could the officer prove that the average of the speeds over many tiny subintervals is the overall speed? Write the average as a sum and show that it collapses into $\frac{f[b]-f[a]}{b-a}$.

Once you know that the average of the small subinterval speeds, $\frac{f[x+\delta x]-f[x]}{\delta x}$ for x running from a to $b - \delta x$ in steps of δx is $\frac{f[b]-f[a]}{b-a}$, how do you know that there are adjacent x's with

$$\frac{f[x_{lo} + \delta x] - f[x_{lo}]}{\delta x} \leq \frac{f[b] - f[a]}{b - a} \leq \frac{f[x_{hi} + \delta x] - f[x_{hi}]}{\delta x}$$

with $|x_{lo} - x_{hi}| = \delta x$?
Why is $f'[x_{lo}] \approx f'[x_{hi}] \approx \frac{f[b]-f[a]}{b-a}$?

One way or the other, complete the officer's argument that there was a time when you were going 74. This is the main task of the project.

20.3. The Mean Value Theorem for Pointwise Derivatives

In the Mathematical Background, we state the Mean Value Theorem for Derivatives in its ultimate generality, only assuming weakly approximating pointwise derivatives and those only at interior points. This complicates the proof but is the key in the Math Background to seeing why regular derivatives and pointwise derivatives are the same when the pointwise derivative is continuous.

20.4. Overall Speed IS an Average

Why do we call $(f[b] - f[a])/(b-a)$ an average of $f'[x]$?
First, if $h[x]$ is a continuous function on a interval $[a, b]$, we show that

$$\text{Ave}[h] = \frac{1}{b-a} \int_a^b h[x]\, dx$$

Break the interval $[a, b]$ into n tiny subintervals, a, $a + \delta x$, $a + 2\delta x$, ..., $b - \delta x$ of length $\delta x = (b-a)/n$. Since $h[x]$ is continuous, $h[x] \approx h[\xi]$ for any $\xi \approx x$, and the average of any one value chosen from each of the subintervals is approximately

$$\begin{aligned}
\text{Ave}[h] &\approx \frac{1}{n}(h[a] + h[a + \delta x] + h[a + 2\delta x] + \ldots + h[b - \delta x]) \\
&\approx \frac{b-a}{b-a}\frac{1}{n}(h[a] + h[a + \delta x] + h[a + 2\delta x] + \ldots + h[b - \delta x]) \\
&\approx \frac{1}{b-a}\frac{b-a}{n}(h[a] + h[a + \delta x] + h[a + 2\delta x] + \ldots + h[b - \delta x]) \\
&\approx \frac{1}{b-a} \delta x (h[a] + h[a + \delta x] + h[a + 2\delta x] + \ldots + h[b - \delta x]) \\
&\approx \frac{1}{b-a}(h[a]\,\delta x + h[a + \delta x]\,\delta x + h[a + 2\delta x]\,\delta x + \ldots + h[b - \delta x]\,\delta x) \\
&\approx \frac{1}{b-a} \int_a^b h[x]\, dx
\end{aligned}$$

HINT 20.8. *Why is $\frac{1}{b-a}(h[a]\,\delta x + h[a+\delta x]\,\delta x + h[a+2\delta x]\,\delta x + \ldots + h[b-\delta x]\,\delta x) \approx \frac{1}{b-a}\int_a^b h[x]\, dx$?*

HINT 20.9. *Why is $f[b] - f[a] = \int_a^b f'[x]\, dx$?*

Why is the overall speed $\frac{f[b]-f[a]}{b-a}$ the average of the instantaneous speed?

PROJECT 21

Inverse Functions and Their Derivatives

The inverse of a function $y = f[x]$ is the function $x = g[y]$ whose "rule undoes what the rule for f does." For example, if $y = f[x] = x^2$, then $x = g[y] = \sqrt{y}$, at least when $x \geq 0$. These two functions have the same graph if we plot g with its independent variable where the y axis normally goes, rather than plotting the input variable of $g[y]$ on the horizontal scale.

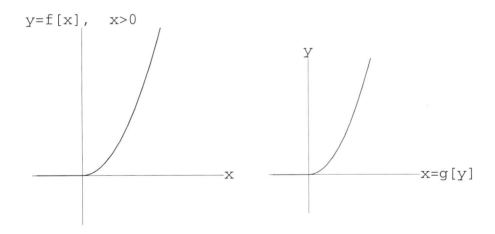

The graph of $x = g[y]$ operationally gives the function g by choosing a y value on the y axis, moving horizontally to the graph, and then moving vertically to the x output on the x axis. This makes it clear graphically that the rule for g "undoes" what the rule for f does. If we first compute $f[x]$ and then substitute that answer into $g[y]$, we end up with the original x. We can also plot both functions with the other orientation of axes.

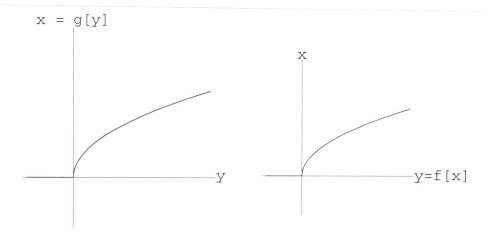

21.1. Graphical Representation of the Inverse

Explain the idea of an inverse function and its graph in your own words.

HINT 21.1. *Reverse or Mirror Inverse*
1) Use general function composition notation to express the statement, "The function g undoes what f does." as an equation. How do you write, "First do f to x, then apply g to the answer." in function notation? What is the answer when you compute this composition? Equate these two things for the answer to this question.

2) If $y = f[x]$ and $x = g[y]$ are inverse functions, $x = g[f[x]]$ and $y = f[g[y]]$, but we plot both functions on the same (x, y) axes, explain why we get the same graph for both functions. How is g "computed" from its graph? In other words, what is the geometric interpretation of "The function g undoes what f does." ?

We can take the usual convention with input variable on the horizontal axis for both functions. This makes the graph of the inverse function the "mirror image" across $y = x$. For this project you will find it best to plot both functions with the x and y axes in the same place.

EXAMPLE 21.2. *A Partial Inverse by Algebra*

Often we can find an inverse function or partial inverse function by solving equations. For example, if $y = x^2 + 1$, then $x^2 = y - 1$ and $x = \sqrt{y-1}$ when $y \geq 1$ and $x \geq 0$. The negative square root also satisfies the original equation, $x = -\sqrt{y-1}$.

The algebra of inverse functions can be tricky, but the calculus of their derivatives is much easier - just look in the infinitesimal microscope.

21.2. The Derivative of the Inverse

HINT 21.3. *The Inverse Function Rule*

(1) Differentiate both sides of the general equation $x = g[f[x]]$ and thus use the Chain Rule to explain symbolically that if $y = f[x]$ and $x = g[y]$ have the same graphs and if both derivatives $f'[x]$ and $g'[y]$ are nonzero, then

$$g'[y] = \frac{dx}{dy} = 1 \Big/ \frac{dy}{dx} = \frac{1}{f'[x]}$$

at corresponding (x, y) pairs.

(2) Why is the microscopic view of the graph $x = g[y]$ the same set of points as the microscopic view of $y = f[x]$ when f and g are inverse functions? (Assuming we keep the y axis vertical to plot both.)

(3) Explain geometrically why the slope of the curve $y = f[x]$, which is the same curve as $x = g[y]$, is the reciprocal of the slope of $x = g[y]$ when you view the vertical y axis as the independent variable for g and the dependent variable for f. What do you see in an infinitesimal microscope?

(4) Even if you did not know that there was an inverse function $x = g[y]$, why does your view in the infinitesimal microscope of $y = f[x]$ convince you that there must be one, at least on a small interval? How does Bolzano's Intermediate Value Theorem 20.2 in the Mean Value Math Police project (or the Math Background) help? What goes wrong if $f'[x] = 0$?

EXAMPLE 21.4. *The Derivative of the Inverse*

It is sometimes easier to compute the derivative of the inverse function and invert for the derivative of the function itself. For example, if $y = x^2 + 1$ and $x = \sqrt{y-1}$ when $y \geq 1$, then $\frac{dy}{dx} = 2x$. The inverse function rule says

$$\frac{dx}{dy} = 1 \Big/ \frac{dy}{dx} = \frac{1}{2x}$$

$$= \frac{1}{2\sqrt{y-1}}$$

We converted the expression $\frac{1}{2x}$ into an expression in the independent variable for g at the last step. We can compute $\frac{dx}{dy}$ directly from rules using the formula $x = \sqrt{y-1}$ to see that these rules agree. Since $x = (y-1)^{\frac{1}{2}}$, the Power Rule and Chain Rule give

$$\frac{dx}{dy} = \frac{1}{2}(y-1)^{\frac{1}{2}-1}$$

$$= \frac{1}{2}(y-1)^{-\frac{1}{2}}$$

$$= \frac{1}{2\sqrt{y-1}}$$

Notice that the first computation is somewhat easier but used conversion of variables at the last step.

EXAMPLE 21.5. *Derivative of Log by Inverse Functions*

The derivative of $x = \text{Log}[y]$ follows from the Inverse Function Rule and the rule for the natural exponential. We have only given these rules without proof, but we observe here that we only need to prove one of the two. The inverse of $x = \text{Log}[y]$ is $y = e^x$ and has derivative $\frac{dy}{dx} = e^x = y$, therefore

$$\frac{d(\text{Log}[y])}{dy} = 1/\frac{dy}{dx}$$
$$= 1/e^x$$
$$= \frac{1}{y}$$

Here is some practice at using the formula for the derivative of the inverse function. Conversion of variables will be a little more trouble with trig functions, but how else would you find the derivative?

HINT 21.6. $\frac{d\,\text{ArcTan}[y]}{dy}$

We know that if $y = \text{Tan}[x]$, then $\frac{dy}{dx} = \frac{1}{(\text{Cos}[x])^2}$. (Hint: $y = \text{Sin}[x](\text{Cos}[x])^{-1}$, so you can use the Chain Rule and Product Rule. Also see text Section 5.3 on CD.) Use the Inverse Function Rule to compute

$$\frac{d(\text{ArcTan}[y])}{dy} = 1/(\frac{dy}{dx})$$

when $y = \text{Tan}[x]$. We want you to express your answer in terms of the independent variable for the arctangent, y, so you need to use identities to convert

$$\frac{dx}{dy} = (\text{Cos}[x])^2$$

into a function of y. We know that $\text{Tan}^2[x] = \frac{\text{Sin}^2[x]}{(\text{Cos}[x])^2} = y^2$ and $\text{Sin}^2[x] + (\text{Cos}[x])^2 = 1$, so we can express $(\text{Cos}[x])^2$ in terms of y.

The Inverse Function Rule can be used numerically even when the formula for the inverse function is not known. The next exercise shows you what we mean.

HINT 21.7. *Graph $y = f[x] = x^5 + x^3 + x$ (use the computer if you wish, but the slope information is important). Use your graph to argue that there must be a function $x = g[y]$ defined for all real y such that $y = f[x]$ if and only if $x = g[y]$. What features of a graph $y = f[x]$ would make g only defined partially (or only on an interval)? (For contrast consider the graphs $y = x^4$ and $y = \frac{1}{1+x^2}$.) How could Bolzano's Intermediate Value Theorem 20.2 in the Mean Value Math Police project (or the Math Background) for f tell you g exists? Why can't you find a formula for g?*

Show that $g'[3] = \frac{1}{9}$ by using numerics on the expression for $g'[y]$ in terms of its dependent variable. (Note that $f[1] = 3$.)

The important Inverse Function Theorem that says that if a function has a non-zero derivative, then at least over an interval, the curve $y = f[x]$ has an inverse function with the same graph (when the same axes are used for both plots). The "proof" (as opposed to the rule that assumes you know g) uses the view in an infinitesimal microscope to compute an approximation to the inverse function. You see now that computation of the derivative of the inverse function is obvious (at least in the dependent variable) once you know the derivative of $y = f[x]$, but it may not be so obvious how to compute the inverse function itself: How would *you* compute arctangent or arcsine *yourself*? This project gives an easy method.

EXAMPLE 21.8. *A Nonelementary Inverse*

Some functions do not have expressions for their inverses. Let
$$y = f[x] = x^x$$
This may be written using $x = e^{\text{Log}[x]}$, so $x^x = (e^{\text{Log}[x]})^x = e^{x\,\text{Log}[x]}$, and $f[x]$ has derivative

$$\frac{dy}{dx} = \frac{d(e^{x\,\text{Log}[x]})}{dx}$$
$$= (\text{Log}[x] + x\,\frac{1}{x})e^{x\,\text{Log}[x]}$$
$$= (1 + \text{Log}[x])\,x^x$$

HINT 21.9. *What is the microscopic view of $y = x^x$ at $(x,y) = (1,1)$? After you have answered this question, use the computer to plot the function. The slope table from zero to infinity is down, over up, so it is "clear" that once $1 + \text{Log}[x] > 0$ ($x > 1/e$), the function is increasing and there is an inverse. (Also, there is another inverse from zero until $x = 1/e$.)*
What is $\lim_{x \downarrow 0} x^x = \lim_{x \downarrow 0} e^{x\,\text{Log}[x]} = ?$ The answer shows why people sometimes write $0^0 = 1$.

It is clear graphically from the previous exercise that $y = f[x]$ has an inverse on either the interval $(0, 1/e)$ or $(1/e, \infty)$. It turns out that the inverse function $x = g[y]$ can not be expressed in terms of any of the classical functions. In other words, there is no formula for $g[y]$. (This is similar to the non-elementary integrals in the computer program **SymbolicIntegr**. We can compute them numerically with NIntegrate, but there is no elementary expression for them. Computer algebra systems have a non-elementary function $\omega(x)$ that can be used to express the inverse.) We look in an infinitesimal microscope to approximate inverse functions in general.

EXAMPLE 21.10. *Microscopic Approximation to the Inverse*

Suppose we have a function $y = f[x]$ and know that $f'[x]$ exists on an interval around a point $x = x_0$ and we know the values $y_0 = f[x_0]$ and $m = f'[x_0]$. In a microscope we would see the graph (Figure 21.11)

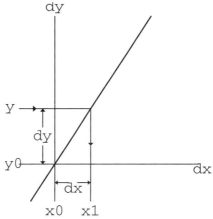
FIGURE 21.11: Small View of $y = f[x]$ at (x_0, y_0)

The point $(dx, dy) = (0, 0)$ in local coordinates is at (x_0, y_0) in regular coordinates.

Suppose we are given y near y_0, $y \approx y_0$. In the microscope, this appears on the dy axis at the local coordinate $dy = y - y_0$. The corresponding dx value is easily computed by inverting the linear approximation

$$dy = m\, dx$$
$$m\, dx = dy$$
$$dx = dy/m$$

What x value, x_1, corresponds to $dx = dy/m$? The answer is $dx = x_1 - x_0$ with x_1 unknown. Solve for the unknown,

$$\begin{aligned} x_1 &= x_0 + dx \\ &= x_0 + dy/m \\ &= x_0 + (y - y_0)/m \end{aligned}$$

Does this value of $x = x_1$ satisfy $y = f[x_1]$ for the value of y we started with? We wouldn't think it would be exact because we computed linearly and we only know the approximation

$$f[x_0 + dx] = f[x_0] + f'[x_0]\, dx + \varepsilon \cdot dx$$
$$f[x_1] = f[x_0] + m \cdot dx + \varepsilon \cdot dx$$

We know that the error $\varepsilon \approx 0$ is small when $dx \approx 0$ is small, so we would have to move the microscope to see the error. Moving along the tangent line until we are centered over x_1, we might see Figure 21.12.

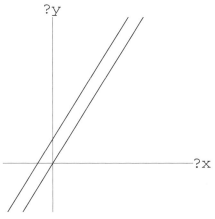

FIGURE 21.12: Small View at (x_1, y)

The graph of $y = f[x]$ appears to be the parallel line above the tangent, because we have only moved x a small amount and $f'[x]$ is continuous by Theorem CD 5.3. We don't know how to compute $x = g[y]$ necessarily, but we do know how to compute $y_1 = f[x_1]$. Suppose we have computed this and focus our microscope at (x_1, y_1) seeing Figure 21.13.

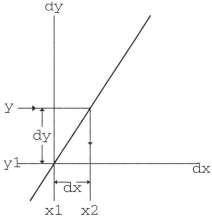

FIGURE 21.13: Small View at (x_1, y_1)

We still have the original $y \approx y_1$ and thus y still appears on the new view at $dy = y - y_1$. The corresponding dx value is easily computed by inverting the linear approximation

$$dy = m\, dx$$
$$m\, dx = dy$$
$$dx = dy/m$$

What x value, x_2, corresponds to $dx = dy/m$? The answer is $dx = x_2 - x_1$ with x_2

unknown. Solve for the unknown,

$$x_2 = x_1 + dx$$
$$= x_1 + dy/m$$
$$= x_1 + (y - y_1)/m$$

This is the same computation we did above to go from x_0 to x_1; in fact, this gives a discrete dynamical system of successive approximations

$$x_0 = \text{given}$$
$$x_{n+1} = x_n + (y - f[x_n])/m$$
$$x_{n+1} = G[x_n], \quad \text{with} \quad G[x] = x + (y - f[x])/m$$

HINT 21.14. *Two Basic Examples*

We know that $y_0 = \text{Sin}[x_0]$ at $(x_0, y_0) = (\pi/6, 1/2) \approx (0.523599, 0.5)$. We also know that $m = \text{Cos}[\pi/6] = \sqrt{3}/2 \approx 0.866025$. Compute a few iterates of the approximation procedure above for $\text{ArcSin}[0.55]$,

$$x_1 = x_0 + (y - f[x_0])/m$$
$$x_1 = 0.523599 + (0.55 - 0.5)/0.866025 = 0.581334$$
$$x_2 = x_1 + (y - f[x_1])/m$$
$$x_2 = 0.581334 + (0.55 - \text{Sin}[0.581334])/0.866025$$
$$= 0.582328$$

$$\vdots$$

and check your work with the computer program **InverseFctHelp** on our website.

Modify the program to compute arctangent, but without using the built-in function. Take $y = \text{Tan}[x]$, $(x_0, y_0) = (0, 0)$. Use the built-in arctangent to check your work. This is how you could compute arctangent yourself.

You will see in the computer program that the approximation for the arcsine converges very rapidly for nearby values of $y \approx 0.5$.

HINT 21.15. *Limits of Approximation*

What is a value of y not near $y = 0.5$ for which you do not expect the approximation procedure for arcsine to converge? (Hint: When is $\text{Sin}[x] = 2$?)

Experiment with the program **InverseFctHelp** to see how far away from 0.5 you can take y and still successfully compute arcsine.

Experiment with the program **InverseFctHelp** to see how far away from 0 you can take y and still successfully compute arctangent.

The general successive approximation scheme to compute the inverse function $x = g[y]$ is

$$x_0 = \text{given}$$
$$x_{n+1} = x_n + (y - f[x_n])/m$$
$$x_{n+1} = G[x_n]$$

We want to show that this tends toward the right value
$$\lim_{n\to\infty} x_n = x_\infty \quad \text{with} \quad y = f[x_\infty]$$

HINT 21.16. *Equilibrium and Inverse*
Prove that $x_\infty = g[y]$ for the inverse function g corresponding to f if and only if x_∞ is an equilibrium point of the dynamical system $x_{n+1} = G[x_n]$ above.

In order for the $\lim_{n\to\infty} x_n = x_\infty$, we need to have a stable dynamical system.

HINT 21.17. *Stability*
If you know that there is an equilibrium point $x_\infty \approx x_0$, apply the Nonlinear Stability Theorem of text Chapter CD 20 to show that it is an attractor. What do you need to know about $f'(x_\infty)$ compared with $f'[x_0]$ for stability?

21.3. Nonelementary Inversion

Conclude your project by computing the inverse function $x = g[y]$ when $y = f[x] = x^x$. Explain why your the computer program is a convergent approximation procedure (even though there is no elementary expression for g).

There is more information on the Inverse Function Theorem in the Mathematical Background book on CD.

PROJECT 22

Taylor's formula

Taylor's formula is a more accurate local formula than the "microscope approximation." It has many uses. Taylor's formula uses the first derivative $f'[x]$ and a number of higher order derivatives: the second derivative $f''[x]$, which is the derivative of $f'[x]$; the third derivative $f^{(3)}[x]$, which is the derivative of $f''[x]$; .... Here is the general result:

THEOREM 22.1. *Taylor's Small Oh Formula*
Suppose that $f[x]$ has n ordinary continuous derivatives on the interval (a,b). If x is not near a or b and $\delta x \approx 0$ is small, then

$$f[x+\delta x] = f[x] + f'[x] \cdot \delta x + \frac{1}{2}f''[x] \cdot \delta x^2 + \frac{1}{3 \cdot 2}f^{(3)}[x] \cdot \delta x^3 + \cdots + \frac{1}{n!}f^{(n)}[x] \cdot \delta x^n + \varepsilon \cdot \delta x^n$$

for $\varepsilon \approx 0$.

When $n = 1$, Taylor's formula is the "microscope approximation,"

$$f[x+\delta x] = f[x] + f'[x] \cdot \delta x + \varepsilon \cdot \delta x$$

Notice that if $\delta x \approx 0$ is small, then its square $(\delta x)^2 = \delta x^2$ is small even on a scale of δx. This makes the second-order formula a more accurate approximation for $f[x+\delta x]$,

$$f[x+\delta x] = f[x] + f'[x] \cdot \delta x + \frac{1}{2}f''[x] \cdot \delta x^2 + \varepsilon \cdot \delta x^2$$

For example, if $\delta x = 1/1000$, then δx^2 is only one one-thousandth of that, $\delta x^2 = 10^{-6}$.

Strictly speaking, the "ε" in the two formulas cannot be compared. All we know is that when δx is "small enough" both ε's are as small as we prescribe. All we really know is that "eventually" the second-order formula is better than the first-order one. In the Mathematical Background Chapter 8, we make the approximation more precise.

HINT 22.2. *Graphical Comparisons*
Show that the Taylor polynomials for sine at $x = 0$ satisfy

$$\text{Sin}[0 + dx] = dx - \frac{1}{6} \cdot dx^3 + \frac{1}{5!} \cdot dx^5 + \cdots$$

and use the computer to compare the plots of several of the approximations,

$$f[dx] = \text{Sin}[0 + dx]$$
$$g[dx] = dx$$
$$h[dx] = dx - dx^3/(3 \cdot 2)$$
$$i[dx] = dx - dx^3/(3 \cdot 2) + dx^5/(5 \cdot 4 \cdot 3 \cdot 2)$$

Make similar graphical comparisons for $\text{Cos}[0 + dx]$ and $\text{Exp}[0 + dx] = e^{0+dx}$.

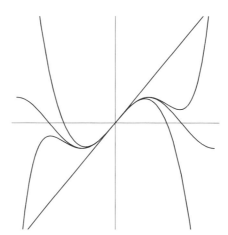

FIGURE 22.3: Sine and 3 Taylor Approximations

22.1. The Increment Equation and Increasing

It is "clear" that if we view a graph in an powerful microscope and see the graph as indistinguishable from an upward-sloping line at a point x_0, then the function must be "increasing" near x_0. The previous Project 21 on inverse functions uses this idea in a very computational way. Certainly, the graph need not be increasing everywhere — draw $y = x^2$ and consider the point $x_0 = 1$ with $f'[1] = 2$. Exactly how should we formulate this? Even if you don't care about the symbolic proof of the algebraic formulation, the formulation itself may be useful in cases where you don't have graphs.

THEOREM 22.4. *Local Monotony*
Suppose the function $f[x]$ is differentiable on the real interval $a < x < b$ and x_0 is a real point in this interval.

(1) *If $f'[x_0] > 0$, then there is a real interval $[\alpha, \beta]$, $a < \alpha < x_0 < \beta < b$, such that $f[x]$ is increasing on $[\alpha, \beta]$, that is,*

$$\alpha \leq x_1 < x_2 \leq \beta \quad \Rightarrow \quad f[x_1] < f[x_2]$$

(2) *If $f'[x_0] < 0$, then there is a real interval $[\alpha, \beta]$, $a < \alpha < x_0 < \beta < b$, such that $f[x]$ is decreasing on $[\alpha, \beta]$, that is,*

$$\alpha \leq x_1 < x_2 \leq \beta \quad \Rightarrow \quad f[x_1] > f[x_2]$$

Math Background Section 5.6 has a more complete exposition of this topic. The idea is simple: Compute the change in $f[x]$ using the positive slope straight line and keep track of the error.

Take x_1 and x_2 so that $x_0 \approx x_1 < x_2 \approx x_0$. Since $f'[x_1] \approx f'[x_0]$ (see text Section CD 5.5 and Math Background Theorem 5.4), we may write $f'[x_1] = m + \varepsilon_1$ where $m = f'[x_0]$ and $\varepsilon_1 \approx 0$. Let $\delta x = x_2 - x_1$ so

$$f[x_2] = f[x_1 + \delta x] = f[x_1] + f'[x_1] \cdot \delta x + \varepsilon_2 \cdot \delta x$$
$$= f[x_1] + m \cdot \delta x + (\varepsilon_1 + \varepsilon_2) \cdot \delta x$$

The number m is a real positive number, so $m + \varepsilon_1 + \varepsilon_2 > 0$ and , since $\delta x > 0$, $(m + \varepsilon_1 + \varepsilon_2) \cdot \delta x > 0$. This means $f[x_2] - f[x_1] > 0$ and $f[x_2] > f[x_1]$.

The Math Background shows how to make the approximations precise and thus allow x_1 and x_2 to range in an interval $[\alpha, \beta]$.

22.2. Taylor's formula and Bending

The smile and frown icons of text Chapter 9 are based on a simple intuitive mathematical idea: when the slope of the tangent increases, the curve bends up. We have two questions. (1) How can we formulate bending symbolically? (2) How do we prove that the formulation is true? First things first.

If a curve bends up, it lies above its tangent line. Draw the picture. The tangent line at x_0 has the formula $y = b + m(x - x_0)$ with $b = f[x_0]$ and $m = f'[x_0]$. If the graph lies above the tangent, $f[x_1]$ should be greater than $b + m(x_1 - x_0) = f[x_0] + f'[x_0](x_1 - x_0)$ or

$$f[x_1] > f[x_0] + f'[x_0](x_1 - x_0)$$

This is the answer to question 1, but now we are faced with question 2. The increment approximation says

$$f[x_1] = f[x_0] + f'[x_0](x_1 - x_0) + \varepsilon \cdot (x_1 - x_0)$$

so this direct formulation of "bending up" requires that we show that the whole error $\varepsilon \cdot (x_1 - x_0)$ stays positive for $x_1 \approx x_0$. All we have to work with is the increment approximation for $f'[x]$ and the fact that $f''[x_0] > 0$. A direct proof is not very easy to give – at least we don't know a direct one. The second-order Taylor formula will make this easy.

We have at least formulated the result as follows.

THEOREM 22.5. *Local Bending*
Suppose the function $f[x]$ is twice differentiable on the real interval $a < x < b$ and x_0 is a real point in this interval.

(1) *If $f''[x_0] > 0$, then there is a real interval $[\alpha, \beta]$, $a < \alpha < x_0 < \beta < b$, such that $y = f[x]$ lies above its tangent over $[\alpha, \beta]$, that is,*

$$\alpha \leq x_1 \leq \beta \quad \Rightarrow \quad f[x_1] > f[x_0] + f'[x_0](x_1 - x_0)$$

(2) *If $f''[x_0] < 0$, then there is a real interval $[\alpha, \beta]$, $a < \alpha < x_0 < \beta < b$, such that $y = f[x]$ lies below its tangent over $[\alpha, \beta]$, that is,*

$$\alpha \leq x_1 \leq \beta \quad \Rightarrow \quad f[x_1] < f[x_0] + f'[x_0](x_1 - x_0)$$

We want you to use the second-order Taylor formula to show the algebraic form of the smile icon. If $f[x]$ is twice differentiable on a real interval (a,b), $a < x < b$, and x is not near a or b, then for any small $\delta x \approx 0$

$$f[x + \delta x] = f[x] + f'[x]\,\delta x + \frac{1}{2} f''[x] \cdot (\delta x)^2 + \varepsilon \cdot \delta x^2$$

with $\varepsilon \approx 0$.

HINT 22.6. *The Local Bending Theorem from Taylor's formula*
Suppose that $f''[x_0] > 0$ at the real value x_0. If $x_1 \approx x_0$, substitute $x = x_0$ and $\delta x = x_1 - x_0$ into Taylor's second-order formula to show the local bending formula. Use the fact that $\frac{1}{2}(f''[x_0] + \varepsilon)(x_1 - x_0)^2 > 0$.

Math Background Section 8.1 has a complete exposition of this topic. In particular, it deals with the question of how far away x_1 can be from x_0.

22.3. Symmetric Differences and Taylor's formula

Taylor's formula can also be used to find a formula for second derivatives and to explain why symmetric differences give a more accurate approximation to first derivatives than the formula $[f[x + \delta x] - f[x]]/\delta x$.

Substitute δx and $-\delta x$ into Taylor's Second Order Formula to obtain

$$f[x + \delta x] = f[x] + f'[x]\delta x + \frac{1}{2} f''[x] \cdot \delta x^2 + \varepsilon_1 \delta x^2$$
$$f[x - \delta x] = f[x] - f'[x]\delta x + \frac{1}{2} f''[x] \cdot \delta x^2 + \varepsilon_2 \delta x^2$$

Subtract the two to obtain

$$f[x + \delta x] - f[x - \delta x] = 2 f'[x] \cdot \delta x + (\varepsilon_1 + \varepsilon_2) \cdot \delta x^2$$

HINT 22.7. *Symmetric Difference Error*

Solve the last formula above for $f'[x]$, obtaining

$$\frac{f[x + \delta x] - f[x - \delta x]}{2\delta x}$$

and an error. Why is this formula algebraically a better approximation for $f'[x]$ than the one you obtain by solving the ordinary increment approximation $f[x + \delta x] = f[x] + f'[x]\delta x + \varepsilon_3 \delta x$ for $f'[x] = [f[x + \delta x] - f[x]]/\delta x + \varepsilon_3$? Compare the errors and note the importance of δx being small.

Graphically, the approximation of slope given by the symmetric difference is clearly better on a "typical" graph as illustrated below. A line through the points $(x, f[x])$ and $(x + \delta x, f[x + \delta x])$ is drawn with the tangent at x in one view, while a line through $(x - \delta x, f[x - \delta x])$ and $(x + \delta x, f[x + \delta x])$ is drawn with the tangent at x in the other. The second slope is closer to the slope of the tangent, even though the line does not go through the point of tangency.

a) b)

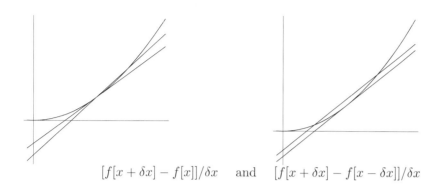

$[f[x+\delta x] - f[x]]/\delta x$ and $[f[x+\delta x] - f[x-\delta x]]/\delta x$

HINT 22.8. *Sketch the line through the points $(x-\delta x, f[x-\delta x])$ and $(x, f[x])$ on the first view. Show that the average of the slopes of the two secant lines on the resulting figure is $[f[x+\delta x] - f[x-\delta x]]/[2\delta x]$, the same as the slope of the symmetric secant line in the second view.*

A quadratic function $q[dx]$ in the local variable dx that matches the graph $y = f[x]$ at the three x values, $x - \delta x$, x, and $x + \delta x$, is given by

$$q[dx] = y_1 + \frac{y_2 - y_1}{\delta x} dx + \frac{y_3 - 2y_1 + y_2}{2\delta x^2}(dx(dx - \delta x))$$

where $y_1 = f[x]$, $y_2 = f[x + \delta x]$, and $y_3 = f[x - \delta x]$.

HINT 22.9. *Verify that the values agree at these points by substituting the values $dx = 0$, $dx = \delta x$ and $dx = -\delta x$.*

Show that the derivative $q'[0] = (f[x + \delta x] - f[x - \delta x])/(2\delta x)$, the same as the symmetric secant line slope. In other words, a quadratic fit gives the same slope approximation as the symmetric one, which is also the same as the average of a left and a right approximation. All these approximations are "second-order."

It is interesting to compare different numerical approximations to the derivative in a difficult, but known case. This is done in Project 17 on direct computation of the derivative of an exponential. The experiments give a concrete form to the error estimates of the previous exercise.

When we only have data (such as in the law of gravity in Chapter 10 of the main text or in the air resistance project in the Scientific Projects), we must use an approximation. In that case the symmetric formula is best.

22.4. Direct Computation of Second Derivatives

Substitute δx and $-\delta x$ into Taylor's second-order formula to obtain

$$f[x + \delta x] = f[x] + f'[x]\delta x + \frac{1}{2}f''[x] \cdot \delta x^2 + \varepsilon_1 \delta x^2$$

$$f[x - \delta x] = f[x] - f'[x]\delta x + \frac{1}{2}f''[x] \cdot \delta x^2 + \varepsilon_2 \delta x^2$$

Add the two to obtain

$$f[x + \delta x] - 2f[x] + f[x - \delta x] = f''[x] \cdot \delta x^2 + (\varepsilon_1 + \varepsilon_2) \delta x^2$$

HINT 22.10. *Second Differences for Second Derivatives*

(1) *The acceleration data in the **Gravity** program from text Chapter 10 is obtained by taking differences of differences. Suppose three x values are $x - \delta x$, x, and $x + \delta x$. Two velocities correspond to the difference quotients*

$$\frac{f[x] - f[x - \delta x]}{\delta x} \quad \text{and} \quad \frac{f[x + \delta x] - f[x]}{\delta x}$$

Compute the difference of these two differences and divide by the correct x step size. What formula do you obtain?

(2) *Solve the equation $f[x + \delta x] - 2f[x] + f[x - \delta x] = f''[x] \cdot \delta x^2 + (\varepsilon_1 + \varepsilon_2) \delta x^2$ for $f''[x]$ and compare this with the answer from part 1 of this exercise. What does the comparison tell you?*

(3) *Use the computer to compute this direct numerical aproximation to the second derivative and compare it with the exact symbolically calculated derivative of several functions. Include some second degree polynomials in yout test functions.*

22.5. Direct Interpretation of Higher Order Derivatives

We know that the first derivative tells us the slope and the second derivative tells us the concavity or convexity (frown or smile), but what do the third, fourth, and higher derivatives tell us?

The symmetric limit interpretation of derivative arose from fitting the curve $y = f[x]$ at the points $x - \delta x$ and $x + \delta x$ and then taking the limit. A more detailed approach to studying higher order properties of the graph is to fit a polynomial to several points and take a limit.

To determine a quadradic fit to a curve, we would need three points, say $x - \delta x$, x, and $x + \delta x$. We would then have three values of the function, $f[x - \delta x]$, $f[x]$, and $f[x + \delta x]$ to use to determine unknown coefficients in the interpolation polynomial $p[dx] = a_0 + a_1 dx + a_2 dx^2$. We could solve for these coefficients in order to make $f[x - \delta x] = p[-\delta x]$, $f[x] = p[0]$, and $f[x + \delta x] = p[\delta x]$. This solution can easily be done with computer commands. The limit of this fit tends to the second-order Taylor polynomial,

$$\lim_{\delta x \to 0} p[dx] = f[x] + f'[x]\, dx + \frac{1}{2} f''[x]\, dx^2$$

This approach extends to as many derivatives as we wish. If we fit to $n+1$ points, we can determine the $n+1$ coefficients in the polynomial
$$p[dx] = a_0 + a_1 \, dx + \cdots + a_n \, dx^n$$
so that $p[\delta x_i] = f[x + \delta x_i]$ for $i = 0, 1, \cdots, n$. If the function $f[x]$ is n times continuously differentiable,
$$\lim_{\delta x \to 0} p[dx] = f[x] + f'[x] \, dx + \frac{1}{2} f''[x] \, dx^2 + \cdots + \frac{1}{n!} f^{(n)}[x] \, dx^n$$
Specifically, if $p[dx] = a_0[x, \delta x] + a_1[x, \delta x] \, dx + \cdots + a_n[x, \delta x] \, dx^n$, then
$$\lim_{\delta x \to 0} a_k[x, \delta x] = \frac{1}{k!} f^{(k)}[x], \qquad \text{for all } k = 0, 1, \cdots, n$$
uniformly for x in compact intervals. The higher derivatives mean no more or less than the coefficients of a local polynomial fit to the function. In other words, once we understand the geometric meaning of the dx^3 coefficient in a cubic polynomial, we can apply that knowledge locally to a thrice differentiable function. This is explored in detail in Math Background Section 8.5.

Chapter 8

Applications to Physics

This chapter begins with three projects related to Newton's law of acceleration,

$$F = m\,a$$

The first considers what happens when a real ladder slides down a wall with the base pulled out at constant velocity. The second project considers air resistance for an object falling a long distance with the approximation of a linear force due to the air. The third project considers nonlinear air resistance and an elastic force – bungee diving.

PROJECT 23

The Falling Ladder (or Dad's Disaster)

Emily's Dad decided to paint the garage. He put a ladder against the wall and climbed to the top. Then the base of the ladder slid out.... How fast was he going when he hit the ground?

In the section on linked variables, or related rates, in Chapter 7 of the core text, we discussed what happens as a ladder slides down a wall. That model is oversimplified in a way that we want you to discover.

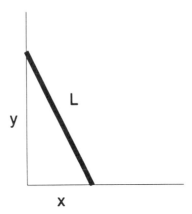

FIGURE 23.1: Ladder & Wall

The simple-minded model suggests that Dad will strike the ground at warp 9, well above the speed of light. This is nonsense. He certainly cannot exceed the speed of light and it seems highly dubious that he would even approach the speed of sound.

Recall the setup for the basic model. The ladder is L feet long. We let x denote the distance from the vertical wall along the horizontal floor to the base of the ladder and y denote the distance up the wall to the place where the ladder rests.

Since $x^2 + y^2 = L^2$, we can solve for y and compute $\frac{dy}{dx} = -\frac{x}{y} = -\frac{x}{\sqrt{L^2 - x^2}}$. By the Chain Rule, $\frac{dy}{dt} = \frac{dy}{dx}\frac{dx}{dt}$ and the speed at which we pull the base away from the wall is constant,

$\frac{dx}{dt} = r$. This predicts that the speed of the tip tends to infinity as x tends to L or y tends to zero, because the denominator of $-r\frac{x}{y}$ tends to zero while the numerator tends to L. (See Section CD7.5 of the main text for more details.)

HINT 23.2. *Prove that we can express Dad's vertical speed $\frac{dy}{dt}$ in several ways*

$$\frac{dy}{dt} = -r\frac{x}{y} = -r\frac{x}{\sqrt{L^2 - x^2}} = -r\frac{\sqrt{L^2 - y^2}}{y}$$

Then use the most convenient expression to answer the following questions.
Suppose the ladder is 20 feet long and we pull the base at the rate of 1 foot per second.

(1) *Find the speed of the tip of the ladder when it is 12 feet above the ground.*
(2) *Find the speed of the tip of the ladder when it is 1 foot above the ground.*
(3) *Find the speed of the tip of the ladder when it is 1 inch above the ground.*
(4) *Find the speed of the tip of the ladder when it is $\frac{1}{1000}$ inches above the ground.*
    Write a computer program and check it by doing part 1 by hand.
    Also use the computer to find roots in the next two parts.
(5) *What is the y position when the model predicts the speed of the tip equals the speed of sound, 1100 ft/sec?*
(6) *What is the y position when the model predicts the speed of the tip equals the speed of light, 186000 mi/sec?*

It should be clear that Dad's speed never exceeds the speed of light. In fact, it seems doubtful that it could exceed the speed of sound. The model is certainly very reasonable when the tip of the ladder is high on the wall, so what goes wrong? This has a simple intuitive explanation, provided you have some intuition for what produces accelerations.

HINT 23.3. *Use the Chain Rule or implicit differentiation to find a formula for Dad's acceleration based on the simple Pythagorean model with $x^2 + y^2 = L^2$. There are several ways to begin, since we want the derivative of $\frac{dy}{dt}$ and have the three formulas above for this quantity. Use $\frac{dx}{dt} = 1$, your formula for $\frac{dy}{dt}$, and simplify your formula as much as posible. Express your answer in terms of y and show that when $x^2 + y^2 = L^2$ and $\frac{dx}{dt} = 1$,*

$$\frac{d^2y}{dt^2} = -\frac{L^2}{y^3}$$

The formula for Dad's acceleration tends to infinity as y tends to the ground, but the only thing to provide acceleration is gravity. (The garage wall can prevent it, but it can't provide it.)

HINT 23.4. *Compute the y value where the acceleration of the tip of the ladder equals -32 ft/sec^2. ($g = 32$ is the value of the gravitational constant in foot-slug-second units.) Compute the y velocity at this point. Compute the amount of time it takes to reach this point if the ladder begins with the tip 16 feet up the wall and the base is pulled out at 1 foot per second.*

The only force acting down on the ladder is gravity. The wall counteracts gravity, but only until the downward acceleration is as much as gravity provides. After that, the ladder simply falls under gravity. Let us begin improving the model by using Galileo's law on Dad. We will ignore the mass of the ladder and its motion, except for it keeping Dad in place

while the ladder rests against the wall. If we let $t = 0$ at the time when the tip of the ladder reaches the acceleration -32 ft/sec^2, the equation for the position of the tip after that time is
$$y = y_0 + v_0 t - 16 t^2$$
(v_0 is negative in our case.) This is the exact solution of Galileo's law
$$\frac{d^2y}{dt^2} = -g$$
when $y = y_0$ and $\frac{dy}{dt} = v_0$ at $t = 0$. This formula neglects air friction on Dad and the ladder.

HINT 23.5. *Verify that if $y[t] = y_0 + v_0 t - 16 t^2$, for constants y_0 and v_0, then $\frac{d^2y}{dt^2} = -g$. How much is g? How much is $y[0]$? How much is $\frac{dy}{dt}[0]$?*

HINT 23.6. *Use the values of y_0 and v_0 that you computed in Hint 23.4 to answer the following:*
  (1) *How much later does the tip of the ladder strike the ground, $t_{crash} = ?$*
  (2) *How far from the wall is the tip when it strikes the ground? Does it still touch?*
  (3) *How fast is the tip traveling when it strikes the ground?*
  (4) *What happens physically to the tip of the ladder when its acceleration reaches -32 ft/sec^2? For example, how far from the wall is the tip at time $t_{crash}/2$?*

A more extensive project would be to revise this model to take the effect of air friction into account. The corresponding computations for the previous exercise become more involved, but you should be able to write down the model in terms of a differential equation with some initial conditions after you work the air resistance project in the next section. We encourage you to explore this extension of the project if it interests you.

23.1. Air Resistance on Dad (Optional)

If we take $t = 0$ at the time when the Dad's acceleration at the tip of the ladder is -32 and account for air resistance in the form of the next project, the differential equations describing his further descent are
$$\frac{dy}{dt} = v$$
$$\frac{dv}{dt} = -32 - bv$$
where $y[0] = y_0$ is the height on the wall where the acceleration of -32 is achieved, and $v[0] = v_0$ is the velocity at this point. The constant b depends on Dad's aerodynamics and would need to be measured.

HINT 23.7. *Read ahead to the air resistance project and write your own explanation of how the differential equations above are derived from Newton's $F = m a$ law.*

A reasonable guess at b in these units is $b = 0.2$. With this information, you can use the **AccDEsol** program to solve the initial value problem with your values of y_0 and v_0.

HINT 23.8. *Modify the AccDEsol program to solve the initial value problem*

$$y[0] = y_0$$
$$v[0] = v_0$$
$$\frac{dy}{dt} = v$$
$$\frac{dv}{dt} = -32 - 0.2v$$

with the position y_0 and velocity v_0 (be careful with signs, this should be negative) of Dad at the point where the ladder reaches an acceleration of -32. Run time to $t = tfinal$ so that $y[tfinal] = 0$, the ground.

How much slower is Dad going because of air resistance?
Why does the air resistance have so little effect in this case?

PROJECT 24

Falling with Air Resistance: Data and a Linear Model

Galileo's law written as a second-order differential equation can be solved to give an explicit formula for the distance an object has fallen. For example, if we start from rest and position zero, measuring down, then

$$s = \frac{g}{2} t^2$$

with $g = 9.8$ in meter-kilogram-second units. (The unit of force is called a "newton" in mks units, so 1 kilogram exerts a gravitational force of 9.8 newtons at sea level on earth.) You may have learned this formula or extensions of it in high school physics. It is interesting because it comes from such a simple law of speeding up, namely that acceleration is constant. It clears up the false intuitive notion that heavy objects fall faster; they don't – the constant g is the same for all objects, at least in vacuum.

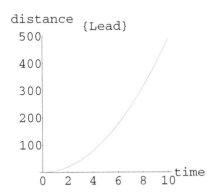 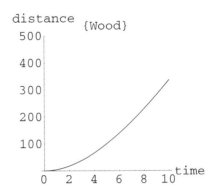

FIGURE 24.1: Free Fall with Air Friction

Light and heavy objects fall at the same speeds if we can neglect air friction. Often we can't neglect that. The computer program **AirResistance** contains data for a wooden ball thrown off the same cliff as the lead ball from the **Gravity** program of Chapter 10 of the main text. We know that a parachute radically affects the speed with which an object falls

- air friction often does matter. The graph of the data from **AirResistance** on our website is shown in Figure 24.1. Notice that it "flattens out" as time increases.

HINT 24.2.

(1) *Compute the velocities $v[t]$ of the object in the data from the AirResistance program from $t = 0$ to $t = \frac{1}{2}$, from $t = \frac{1}{2}$ to $t = 1$, etc. What happens to speed as time increases?*
(2) *Use your velocity list to compute the accelerations $a[t]$ from $t = \frac{1}{2}$ to $t = 1$, from $t = 1$ to $t = \frac{3}{2}$, etc. What happens to acceleration as time increases?*

After Galileo, Newton discovered calculus and derived a more general acceleration law commonly known by

$$F = ma$$

"Force equals mass times acceleration." Newton also related mass to force due to gravity by

$$F = mg$$

so massive objects weigh more but still fall by the same acceleration because m cancels on both sides of

$$ma = mg$$
$$a = g$$
$$\frac{d^2s}{dt^2} = g$$

The differential equation

$$\frac{d^2s}{dt^2} = g, \quad \text{the same constant for all objects}$$

is Galileo's law expressed in the language of calculus (which, hopefully, you observed in text Chapter 10). Newton's law in the language of calculus is

$$m\frac{d^2s}{dt^2} = F, \quad \text{the total applied force}$$

We want to use Newton's law to formulate a more realistic model of a falling body that IS affected by air friction. The new model does not have a simple explicit solution like $s = \frac{g}{2}t^2$. It does have a messy one that you will be able to compute later this semester, but the important thing is the use of calculus as a language describing the changes caused by gravity and air. More complicated models will not have any explicit solutions but still can be described by simple differential equations.

What can we say about air friction? The faster we go, the more the air resists. If you walk and hold your arm out, you hardly notice the resistance of air. If you bike and hold your arm out, you notice, but the force is mild. If you drive on the interstate and stick your arm out the car window at 65 mph, you notice a substantial force. About the simplest law we can formulate with this property is that the air friction force is proportional to the speed,

but we need to keep track of directions. The force from air is opposite to the direction of motion, so
$$F_{air} = -c\frac{ds}{dt}$$
Air friction force is opposite to the direction of the velocity and proportional to its magnitude.

An object falling under gravity has two forces acting on it, gravity pulls down and wind resistance pushes up – if we move,
$$F_{total} = F_{gravity} + F_{air}$$
where the force due to air is negative (because of its direction). We write Newton's "$F = ma$" to get
$$m\frac{d^2s}{dt^2} = mg - c\frac{ds}{dt}$$
divide by the mass m and let the unknown constant $\frac{c}{m} = b$,
$$\frac{d^2s}{dt^2} = g - \frac{c}{m}\frac{ds}{dt}$$
$$\frac{dv}{dt} = g - bv$$
where $v = \frac{ds}{dt}$, the velocity.

HINT 24.3. *Find a constant b so that the model $\frac{dv}{dt} = g - bv$ fits the data in the program* **AirResistance**. *You computed velocity and acceleration in the previous exercise using the computer. You can do list computations to estimate b.*

24.1. Terminal Velocity

There is an interesting way to measure b. If you look carefully at the data for a longer fall, you will see that the wooden ball approaches a limiting velocity. This is just another way to say that the acceleration "flattens out." This limiting velocity is called the terminal velocity, v_T, of the falling object. Unfortunately, we would have to drop the ball out of an airplane to get enough data. (The data we have do not flatten out all the way, the final accelerations are not quite zero.)

The differential equations and initial conditions describing the motion of the ball are
$$y[0] = 0$$
$$v[0] = 0$$
$$\frac{dy}{dt} = v$$
$$\frac{dv}{dt} = g - bv$$

EXPLORATION 24.4. *Terminal Computations*

(1) *If $v[t]$ tends to a limit, then its rate of change becomes less and less as it does so. Why?*

(2) What is the calculus expression for the time rate of change of v? According to part (1), this expression tends to zero as the velocity tends to its limiting value.
(3) At terminal velocity, v_T, $\frac{dv}{dt} = 0$. Use the differential equation for v to find v_T in terms of g and b.
(4) What is the terminal velocity of the wooden ball in air? In other words, how fast would you expect it to be going when it hit the ground if dropped from an airplane? Hint: Use your computation of b from the data.

24.2. Comparison with the Symbolic Solution

A trick called "separation of variables" that we will learn later shows that the solution to the air resistance initial value problem is

$$y[t] = \frac{g}{b}\left(t + \frac{1}{b}\left[e^{-bt} - 1\right]\right)$$

HINT 24.5. *Matching Theory and Data*
1) Show that $y[0] = 0$.
2) Use symbolic rules of calculus to show that $v[t] = \frac{dy}{dt} = \frac{g}{b}\left(1 - e^{-bt}\right)$.
3) Show that $v[0] = 0$.
4) Show that $\frac{dv}{dt} = g e^{-bt}$ and that $g - bv[t] = g e^{-bt}$ also, so that $\frac{dv}{dt} = g - bv$.

Now wrap up your whole project. Explain why $y[t] = \frac{g}{b}\left(t + \frac{1}{b}\left[e^{-bt} - 1\right]\right)$ is the symbolic solution to the differential equations describing the acceleration due to gravity and air resistance. Then use your measured value of b and plot $y[t]$ together with the data from the **AirResistance** program.

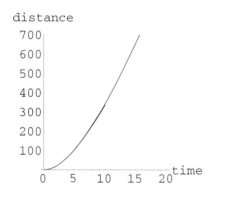
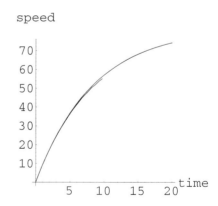

FIGURE 24.6: Comparison of Symbolic Solution and Data

PROJECT 25

Bungee Diving

The Dangerous Sports Club, founded in 1977, periodically embarks on expeditions in which they participate in unusual, exciting, and frequently life-threatening activities. In this project, we will work through one of their unusual activities, bridge jumping. A bridge jumping participant attaches one end of a bungee cord to himself and the other end to the bridge. He then dives off the bridge, hoping he has correctly calculated the length of the cord and that it pulls him up before he hits the bottom of the canyon. One jump took place at the Royal Gorge bridge, a suspension bridge spanning the 1053-foot deep Royal Gorge in Colorado. Two jumpers used 120-foot cords, two others used 240 feet cords, and the last jumper used a 415-foot cord in hopes of touching the bottom of the canyon. Krazy Keith tried this later with a 595-foot cord.

Your job in this project is to find out which jumpers lived. How far did each one fall before the cord started to pull him back up? How far below the bridge did he come to rest? How hard did the cord yank on his leg? Did he survive or did he calculate the length or strength of the cord incorrectly?

Newton's law "$F = m\,a$" says the acceleration of our jumper is proportional to the total force on him. Gravity produces a constant downward force. If the bungee cord is stretched past its natural relaxed length, it pulls up. Of course, this force is the lifesaver – unless it's too weak or too strong. A long fall before stretching the cord results in high speed, and air resists high-speed motion. This is the third force on the jumper. The three forces together with Newton's law tell us the jumper's fate.

25.1. Forces Acting on the Jumper before the Cord Is Stretched

Until the jumper is L feet below the bridge, where L is the length of the cord, the bungee cord exerts no force on the jumper. The only forces we need to consider before the jumper is L feet below the bridge are the force of gravity and air resistance. According to Newton's law, the force of gravity is

$$F_g = m\,g = w = 32\,m$$

where

$$F_g = \text{the force of gravity in pounds}$$
$$m = \text{the mass of the jumper in slugs}$$
$$w = \text{the weight of the jumper in pounds}$$
$$m = \frac{w}{32}$$

The force of gravity is equal to the jumper's weight, which for this project we will assume is 160 pounds. In other words, his mass is 5 slugs. This means the force of gravity is constantly 160 pounds (s. ft./.sec.^{2}) acting in the downward direction.

The force of air resistance (F_r), on the other hand, is always changing with velocity. The force of air resistance is opposite to velocity (just as in the air resistance project), but proportional to the $\frac{7}{5}$ power, because the diver goes head down with his arms tucked in for maximum thrill.

$$F_r = -k\,v^{7/5}$$

where

$$F_r = \text{the force of air resistance}$$
$$v = \text{the velocity of the jumper}$$
$$k = \text{the proportionality constant of air resistance}$$

The proportionality constant in this case is 0.1.

HINT 25.1. *You should verify that at 100 ft/sec = 68 mph, there is a 63-pound force on the diver due to air resistance. How much is the force at 133 mph (not 133 ft/sec)?*

VARIABLES

There are an awful lot of letters flapping in the breeze. Let's settle on some basic variables:

$$h = \text{the height of the jumper above the canyon floor, in feet}$$
$$t = \text{the time, in seconds, measured so that } t = 0 \text{ when he jumps}$$

In terms of these variables, we have $h[0] = H_0 = 1053$, the height of the bridge. The velocity is the time derivative of height,

$$\frac{dh}{dt} = v \quad \text{and acceleration is the rate of change of velocity} \quad a = \frac{dv}{dt} = \frac{d^2h}{dt^2}$$

We write Newton's law

$$m\,a = F$$
$$m\frac{d^2h}{dt^2} = F_g + F_r$$
$$m\frac{dv}{dt} = -m\,g - kv^{7/5} \quad \text{so} \quad \frac{dv}{dt} = -g - \frac{k}{m}v^{7/5}$$

with $g = 32$ and $k = 0.1$. Before the bungee cord is taut we have the initial value problem

$$h[0] = H_0 = 1053$$
$$v[0] = 0 \quad \text{he starts from rest}$$
$$\frac{dh}{dt} = v$$
$$\frac{dv}{dt} = -32 - cv^{7/5}, \qquad c = \frac{k}{m}$$

HINT 25.2. *Use the **BungeeHelp** program on our website to solve the above initial value problem, just as we solved for s and i in the **SIRsolver** program in Chapter 2 of the text. We have solved for 5 seconds in the help program and found a speed of 121 ft/sec downward. You need to find the speeds when the three lengths of bungees tighten, $h = 1053 - L$, $L = 120$, $L = 240$, $L = 415$. Adjust the final time of solution until you get these distances.*

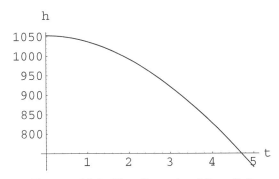

FIGURE 25.3: Five Seconds of Free Fall

HINT 25.4. *If the cord becomes un-attached, how fast will the diver be going when he hits the bottom of the canyon? What velocity makes $\frac{dv}{dt} = 0$? Give your answer in ft/sec and mph.*

25.2. Forces Acting on the Jumper after He Falls L Feet

After the jumper falls L feet below the bridge, the cord is being stretched past its natural position and a third force acts on the jumper, the force of the cord (F_c). The force acting to restore the cord to its natural position is proportional to the amount the cord is stretched past its natural position and acts in the direction that restores it to its natural position.

$$F_c = s\,d$$

where

F_c = the force acting to restore the cord to its natural position
s = the spring constant
d = the amount the cord is stretched past its natural position

The spring constant for this model will be 3.4 for the 120 ft cord. This is equivalent to saying that for every one foot the cord is stretched past its natural position, it exerts a force of 3.4 pounds in the opposite direction. Longer cords are "stretchier", $s = 3.4 \times 120/L$, for natural length L.

HINT 25.5. *Express the distance that the bungee cord is stretched in terms of L, its natural length, and the basic variable h. Use this to give a formula for the force from the bungee in terms of h and the constant s*

$$F_c = s(\ ?\ ) \quad \text{if } h < 1053 - L$$
$$= 0 \quad \text{if } h > 1053 - L$$

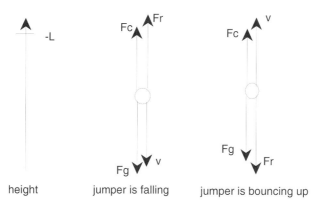

FIGURE 25.6: Forces Acting on Bridge Jumper

HINT 25.7. *With no dynamics, how much would you have to stretch the bungee cord in order to produce a force of 160 pounds? At what height above the canyon will the various divers come to rest? Test your formula for $F_c[h]$ by plugging in your answer to the second question and checking to be sure your formula gives 160.*

The computer can give a piecewise function using the If[.] command. This is illustrated in the **BungeeHelp** program. Once you write a formula for the cord force strictly in terms of h (and the parameter $s = 3.4$), $F_c = F_c[h]$, this can be used in Newton's law to find a model of the diver on the cord.

$$m\, a = F_g + F_r + F_c$$
$$= -m\, 32 - kv^{7/5} + F_c[h]$$

HINT 25.8. *Use Newton's "$F = m\, a$" law with all the forces to write the equations for the jumper as an initial value problem of the form:*

$$h[0] = ?$$
$$v[0] = ?$$
$$\frac{dh}{dt} = v$$
$$\frac{dv}{dt} = ??$$

25.3. Modeling the Jump

Now you are ready to use the differential equation solver in **BungeeHelp** to answer:

HINT 25.9. *The big questions:*

(1) *How far above the canyon floor is the diver at his closest approach?*
(2) *What if h_{min} is negative?*
(3) *How fast is the diver going at his fastest speed? Is it up or down?*
(4) *What is the greatest force on the diver's leg? Can a human leg withstand that force?*
(5) *Did he live? (Remember that Krazy Keith is 5 feet 8 inches tall and weighs only 140 pounds.)*

PROJECT 26

Planck's Radiation Law

Hot bodies emit radiation. No, this isn't the start of a racy novel; we're referring to a blacksmith shop where iron comes out of the fire "white hot," cools to "red hot," and passes below the visible spectrum into the infrared. (Warm human bodies do emit infrared radiation, which is detected by the "motion" lights many of us have outside our houses.) Physicists wanted to be able to predict the "color" of radiation from the temperature.

Wien made measurements and found an empirical law that says: "When the absolute temperature of a radiating black body increases, the most powerful wavelength decreases in such a way that

$$\lambda_{\max} T = W$$

is constant." The measured constant $W \approx 0.28978$ (cm°K) is known as Wien's displacement constant.

Planck discovered a formula for the intensity of each frequency radiating from a black body. When we maximize intensity as a function of wavelength using Planck's formula we can derive Wien's law. This derivation was the first triumph of quantum mechanics. The classical theory made a wrong prediction and the new theory gave the right one and shows why Wien's law holds. We will show how Planck's law gives a theoretical derivation of Wien's law and how calculus, the computer and graphing interact in seeing this. We think this should show you why you should learn the rest of the chapter.

Planck began his study with a graph of the measured intensities, an empirical curve. He then fit a formula to this curve in what his 1920 Nobel Prize lecture described as "an interpolation formula which resulted from a lucky guess." The lucky guess led him to a derivation from physical laws. If you go to your local quantum mechanic's shop, just next door to the blacksmith, she will tell you that the intensity of radiation for the frequencies between ω and $\omega + d\omega$ of a black body at absolute temperature T is

$$I\, d\omega = \frac{\hbar \omega^3}{\pi^2 c^2 (e^{\hbar \omega/(kT)} - 1)} d\omega$$

where $\hbar = \frac{h}{2\pi}$ and $h \approx 6.6255 \times 10^{-27}$ (erg sec) is called Planck's constant, $c \approx 2.9979 \times 10^{10}$ (cm/sec) is the speed of light, and $k \approx 1.3805 \times 10^{-16}$ (erg/deg) is called Boltzmann's constant. Physicists treat this formula as a differential because you need to measure a small range of frequencies, not just the intensity at one frequency.

26.1. The Derivation of Planck's law of Radiation

We want to be sloppy about the units in this example and just emphasize the difficulty we encounter because of the scales of the real physical constants. Normally, we will not accept this practice ourselves or as part of your homework, but we don't want to get side-tracked with the physics. The *Feynman Lectures on Physics* derive Planck's law (in angular frequency form) with their usual mixture of brilliance, charm and mathematical giant steps (flim-flam?). We won't repeat that. If you wish to look that up, you can also straighten out the units and give careful definitions of the variables.

26.2. Wavelength Form and First Plots

We want to express Planck's law in terms of wavelength λ using the basic relation

$$\omega \lambda = 2\pi c$$

and its differential, so

$$\omega = \frac{2\pi c}{\lambda} \quad \text{and} \quad d\omega = -\frac{2\pi c}{\lambda^2} d\lambda$$

giving

$$I\, d\lambda = 8\pi hc^2 \frac{1}{\lambda^5 (e^{hc/(k\lambda T)} - 1)} d\lambda$$

$$= a \frac{1}{\lambda^5} \frac{1}{e^{b/(\lambda T)} - 1} d\lambda$$

where $a = 8\pi hc^2$ and $b = hc/k$.

Let's take a hot object, say $T = 1000°$K (about 1300° F). The first lesson the computer learns from hand computations is what range to plot. Normally, in a textbook problem to plot a function like $i = a/\lambda^5$, you might pick a range like $0 < \lambda < 10$. This is nonsense in our problem, because the interesting range of wavelengths is orders of magnitude smaller. Let's try it with *Mathematica* in Figure 26.1 anyway:

$$a = 1.4966 \; 10^{-4}$$
$$b = 1.4388$$
$$T = 10^3$$
$$i := a/(l^5(Exp[b/(l\;T)] - 1))$$
$$i$$
$$Plot[i, \{l, 0, 10\}]$$

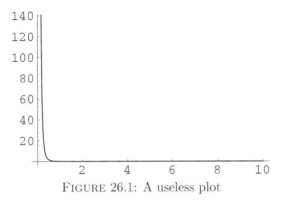

FIGURE 26.1: A useless plot

The wavelength of red cadmium light in air is used for calibrations and equals

$$\lambda = 6438\text{A} = 6438 \times 10^{-8}\text{cm} = 6.438 \times 10^{-5}\text{cm}$$

so let's try another range in Figure 26.2.

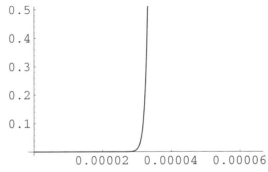

FIGURE 26.2: Also useless

This plot tells us something, at least. The interesting stuff must be between these ranges, so we try a range of $\{l, 0, .002\}$ in Figure 26.3.

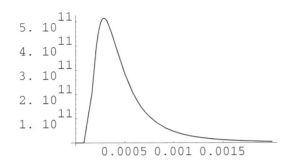

FIGURE 26.3: Intensity for $0 < \lambda < .002$

Now we see what we want – the maximum. It seems to be around 0.0003 as well as we can read the graph. We didn't really use calculus, but a little physics to look for a reasonable range. We were lucky to find the peak. What we wanted was the place where the curve peaks, so we could have used calculus to find it. And we can do that now. Let $B = \frac{b}{T}$, so $I = a\lambda^{-5}(e^{B/\lambda} - 1)^{-1}$ and use the Product Rule and Chain Rule:

$$\frac{dI}{d\lambda} = a\left(-5\lambda^{-6}(e^{B/\lambda} - 1)^{-1} + \lambda^{-5}(-1)(e^{B/\lambda} - 1)^{-2}(B(-\lambda^{-2}e^{B/\lambda})\right)$$

$$= \frac{a}{\lambda^6} \cdot \frac{5 - 5e^{b/\lambda T} + \frac{b}{\lambda T}e^{b/\lambda T}}{(e^{b/\lambda T} - 1)^2}$$

$$= \frac{1.4966 \ 10^{-4}}{\lambda^6} \cdot \frac{\left(5 - 5e^{1.4388 \ 10^{-3}/\lambda} + \frac{1.4388 \ 10^{-3}}{\lambda}e^{1.4388 \ 10^{-3}/\lambda}\right)}{(e^{1.4388 \ 10^{-3}/\lambda} - 1)^2}$$

It is horribly messy to do this computation with the explicit values of the constants. We want you to learn to use parameters, even in cases like this where they only make the calculations neater and thus easier to follow. We will look at the computation with parameters below for a more important reason. Actually, we asked you to differentiate Planck's formula in the last chapter and showed you the answer in the computer program **Dfdx**. All the computations in this section are on the program **PlanckL** and we encourage you to compare that with your Planck's law exercise below.

The graph is horizontal when $\frac{dI}{d\lambda} = 0$ so the numerator must equal zero

$$5 - 5e^{1.4388 \ 10^{-3}/\lambda} + \frac{1.4388 \ 10^{-3}}{\lambda}e^{1.4388 \ 10^{-3}/\lambda} = 0$$

If we try to solve $\frac{dI}{d\lambda} = 0$ using the computer, we encounter nasty numerical problems because of the scale. A change of variables makes the problem appear simpler and scales it in a form that the computer can handle. Let $x = \frac{1.4388 \ 10^{-3}}{\lambda}$ and rewrite the equation above to obtain the problem:

Find x so that $\quad 5 = (5 - x) \ Exp[x]$

We need a starting guess at a root. Figure 26.4 shows a root near $x = 5$.

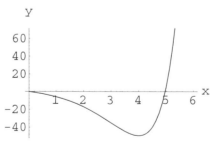

FIGURE 26.4: $y = 5 - (5 - x)Exp[x]$

The computer finds our root more accurately to be $x = \frac{1.4388 \ 10^{-3}}{\lambda} \approx 4.96511$, so $\lambda \approx \frac{1.4388}{4.96511} \times 10^{-3} = 2.89782 \times 10^{-4}$. The observed wavelength at $1000°$ K is 0.000289782 cm. Very nice start, but this is not the answer we are looking for. We have only shown what happens at $T = 1000$ and we want to know what happens as we vary T.

We could try another value, say $T = 900$, and work through the whole problem again, but that would just give us another specific number. We need a formula that tells us how the maximum point depends on the temperature T. We can find this by solving the problem as before, but keeping the letter T as an unknown constant or parameter rather than letting $T = 1000$.

26.3. Maximum Intensity in Terms of a Parameter

The peak on each curve $I = I[\omega]$ where T is constant (but not fixed at some specific number) still occurs the point where the slope is zero or

$$\frac{dI}{d\lambda} = 0$$

We want to solve the equation with the parameter T in the expression. This isn't so hard given all our preliminary work. We know

$$\frac{dI}{d\lambda} = \frac{a}{\lambda^6} \cdot \frac{5 - 5e^{b/\lambda T} + \frac{b}{\lambda T}e^{b/\lambda T}}{(e^{b/\lambda T} - 1)^2}$$

and the derivative is zero if the numerator is zero, or

$$5 = (5 - \frac{b}{\lambda T})e^{\frac{b}{\lambda T}}$$

Substituting $x = \frac{b}{\lambda T}$, we are left with exactly the problem we solved numerically above,

$$\text{Find } x \text{ so that} \quad 5 = (5 - x)e^x$$

which holds when $x \approx 4.96511$.

What do we conclude? At the peak,

$$\frac{b}{\lambda T} \approx 4.96511$$

or

$$\lambda T \approx \frac{1.4388}{4.96511} = 0.289782$$

This is Wien's law, $\lambda_{\max} T = W$ constant. We have shown that the empirical law follows theoretically from Planck's quantum mechanical law and derived $W = 0.289782$ versus the measured $W = 0.28978$. The formulas are a little complicated and calculus plays a role in just finding plot ranges. It plays an indispensable role in finding the *formula* for the peak by differentiating and solving with parameters.

In summary, for fixed temperature, T, the maximum intensity occurs at the wavelength satisfying

$$\lambda_{\max} T = 0.289782$$

or $\lambda_{\max} = 0.289782/T$. The solution to the symbolic maximization problem gives us a function of the parameter T. This is a derivation of Wien's empirical law from Planck's theoretical law.

You can work out the frequency form of Planck's law in **PlanckF** and compare it to the computer wavelength solution just described.

HINT 26.5. *Planck's law written in terms of frequency does not lead to Wien's same constant. We want you to show this by finding the maximum of $I[\omega]$. Use the computer program* **PlanckF** *to help with the computations and sketch curves like the ones we have done for wavelength. The intensity formula with frequency is*

$$I = \frac{\hbar \omega^3}{\pi^2 c^2 (e^{\hbar \omega /(kT)} - 1)}$$

$$= \frac{\alpha \omega^3}{e^{\beta \omega / T} - 1}$$

(1) *Sketch I versus ω on an appropriate range. Figures 26.6 and 26.7 are two poor starting attempts:*

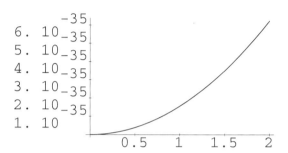

FIGURE 26.6: *A Bad Range of Frequency*

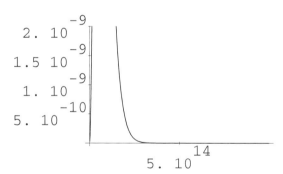

FIGURE 26.7: *A Better Range*

(2) *Compute the derivative*

$$\frac{dI}{d\omega} = a\omega^2 \frac{(3 - \frac{b\omega}{T})e^{\beta \omega/T} - 3}{(e^{\beta \omega/T} - 1)^2}$$

and show that this leads to the properly scaled root-finding problem

$$(3 - x)e^x = 3$$

(3) *Use the computer to solve this equation. (There is help in* **PlanckF**.*)*
(4) *Use the root of the previous part to show that the maximum angular frequency occurs when $\frac{\beta \omega}{T} \approx 2.82144$ so $\lambda T \approx \frac{hc}{k 2.82144} = 0.509951$, since $\lambda = \frac{hc}{\hbar \omega}$.*

(5) *(Optional) Ask your neighborhood quantum mechanic why maximum frequency gives a different Wien's law.*

PROJECT 27

Fermat's principle Implies Snell's law

Snell's law from optics says that light traveling from point P to point Q reflects off a mirror so that the angle of incidence equals the angle of reflection. This is illustrated in Figure 27.1. The angle of incidence is i between the incoming ray of light and the normal to the mirror. The angle of reflection is r between the normal and the outgoing ray of light.

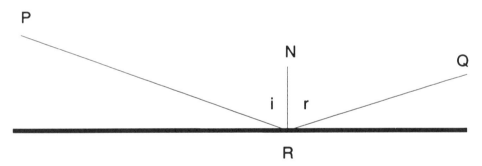

FIGURE 27.1: Snell's law, Angle of Incidence, i = r, Angle of Reflection

Fermat gave a more general variational principle governing the path of light. Fermat's principle says that light travels along the path that requires the least time. Among all paths reflected off a flat mirror, we can see geometrically that light will reflect at the point where the angle of incidence equals the angle of reflection. In other words, Fermat's principle implies Snell's law.

To show that Fermat's principle gives Snell's law, suppose light reflected at a different point S on the mirror. We will show that the time required for light to travel from P to Q through S is more than the time required to travel from P to Q through R, where $i = r$ as shown in Figure 27.1.

Construct the point P' an equal distance below the mirror on a line normal to the mirror. Since the angles i and r are equal, the points P', R, and Q lie on a line. By similar triangles, the distance traveled by light going from P to R is the same as that traveled by light going from P' to R, so the distance traveled from P to R to Q is the length of $P'Q$.

The distance traveled by light going from P to S is the same as that traveled by light going from P' to S by similar triangles, so the distance traveled by light going from P to S and then to Q is the sum of the two lengths $P'S$ plus SQ. Since these segments form the sides of a triangle with $P'Q$ as a base, $P'S + SQ > P'Q$. In other words, P to R to Q is the shortest reflected path.

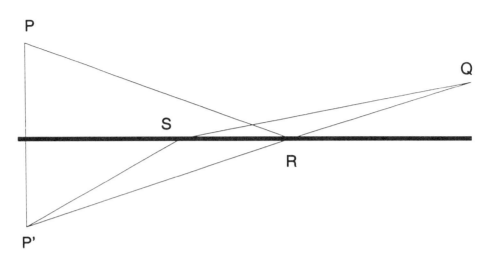

FIGURE 27.2: Proof of Snell's law for a Flat Mirror

Finally, light travels at constant speed c in air, so the shortest path is fastest and Fermat's principle says light must reflect at the point R that makes $i = r$. In other words, Fermat's principle implies Snell's law.

Before we move on to study reflection off curved mirrors, it is worthwhile to give a calculus proof of the shortest path. We need to formulate the problem analytically.

HINT 27.3. *Fermat $\Rightarrow$ Snell by Calculus*
Place coordinates on Figure 27.1 above so that the mirror forms the x-axis. Put P at the point (a, b) and Q at (u, v).

1) Use the Pythagorean theorem to show that the length of the line segment from (a, b) to the reflection point $(x, 0)$ is $L_P = \sqrt{(x - a)^2 + b^2}$.

2) Use the Pythagorean theorem to show that the length of the line segment from $(x, 0)$ to (u, v) is $L_Q = \sqrt{(u - x)^2 + v^2}$.

3) Light travels at speed c, so the time required for light to go from (a, b) to $(x, 0)$ is the distance over c. Similarly, the time required to go from $(x, 0)$ to (u, v) is that distance over c. Show that the total time required for a ray of light to travel this path is

$$t[x] = \frac{1}{c}\left(\sqrt{(x - a)^2 + b^2} + \sqrt{(u - x)^2 + v^2}\right)$$

4) Show that

$$\frac{dt}{dx} = 0 \quad \Leftrightarrow \quad \frac{x - a}{L_P} = \frac{u - x}{L_Q}$$

146

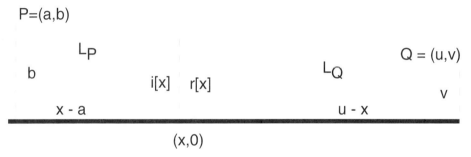

FIGURE 27.4: Coordinates for the Path of Light

5) The angles of incidence and reflection at $(x, 0)$ are shown on Figure 27.4. Prove that

$$\frac{dt}{dx} = 0 \quad \Leftrightarrow \quad i[x] = r[x]$$

5) Prove Snell's law, that the shortest reflected path makes $i[x] = r[x]$. Why is the critical point minimal?

27.1. Reflection off a Curved Mirror

Now we reflect light off a curved mirror. For simplicity, we assume that the mirror is in the shape of a graph $y = f[x]$. A light ray hitting the graph at $(x, f[x])$ is shown in Figure 27.5.

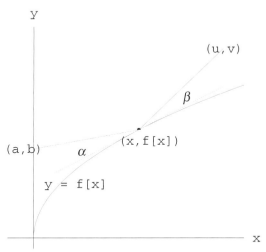

FIGURE 27.5: Reflection off $y = f[x]$

The length $L_P = \sqrt{(x-a)^2 + (f[x]-b)^2}$ and the length $L_Q = \sqrt{(u-x)^2 + (v-f[x])^2}$ (by the length formula or Pythagorean theorem), so the time required for light to pass along the path is

$$t[x] = \frac{1}{c}\left(\sqrt{(x-a)^2 + (f[x]-b)^2} + \sqrt{(u-x)^2 + (v-f[x])^2}\right)$$

To minimize $t[x]$, we begin by computing critical points,

$$c\frac{dt}{dx} = \frac{(x-a) + (f[x]-b)\,f'[x]}{L_P} - \frac{(u-x) + (v-f[x])\,f'[x]}{L_Q}$$

so that critical points satisfy

$$\frac{(x-a) + (f[x]-b)\,f'[x]}{L_P} = \frac{(u-x) + (v-f[x])\,f'[x]}{L_Q}$$

In order to easily interpret this formula, we need the fact that the cosine of the angle between two vectors

$$\mathbf{A} = \begin{bmatrix} \Delta x \\ \Delta y \end{bmatrix} \quad \text{and} \quad \mathbf{T} = \begin{bmatrix} 1 \\ m \end{bmatrix}$$

is given by the simple formula

$$\mathrm{Cos}[\theta] = \frac{\mathbf{A} \bullet \mathbf{T}}{|\mathbf{A}| \cdot |\mathbf{T}|} = \frac{\Delta x \cdot 1 + \Delta y \cdot m}{\sqrt{(\Delta x)^2 + (\Delta y)^2} \cdot \sqrt{m^2 + 1}}$$

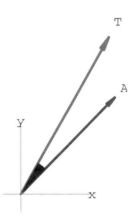

FIGURE 27.6: The angle between $\mathbf{A}$ and $\mathbf{T}$.

You can study this formula in Chapter 15 of the text, or just use it now. Applying this formula to the angles between the tangent vector $\begin{bmatrix} 1 \\ m \end{bmatrix} = \begin{bmatrix} 1 \\ f'[x] \end{bmatrix}$ and each of the vectors from P to $(x, f[x])$, $\begin{bmatrix} (x-a) \\ (f[x]-b) \end{bmatrix}$, and $(x, f[x])$ to Q, $\begin{bmatrix} (u-x) \\ (v-f[x]) \end{bmatrix}$, gives the cosines of the angles,

$$\mathrm{Cos}[\alpha[x]] = \frac{(x-a) + f'[x](f[x]-b)}{L_P\sqrt{1 + (f'[x])^2}} \quad \text{and} \quad \mathrm{Cos}[\beta[x]] = \frac{(u-x) + f'[x](v-f[x])}{L_Q\sqrt{1 + (f'[x])^2}}$$

HINT 27.7. *Use these formulas to show that Fermat's least time principle implies Snell's law of equal angles of incidence and reflection for curved mirrors.*

27.2. Computation of Reflection Angles

Now that we know Snell's law for curves, we can turn the reflection law around and ask for the direction of a ray that reflects off a curve when it is emitted from a given point. In other words, we assume that (a, b) is given and compute the position (u, v) as a function of x. Since we mostly want a direction, we may either take $L_Q = 1$ or change x by 1, $u = x+1$.

In this problem, it helps to use the geometry as an intermediate step, even though our final calculations will all be based on slopes. The connection with geometry is that the slope of the segment from (a, b) to $(x, f[x])$ is the tangent of the angle the segment forms with the horizontal,

$$\text{Tan}[\theta_1] = \frac{f[x] - b}{x - a}$$

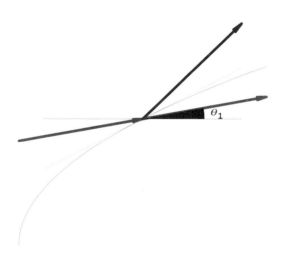

FIGURE 27.8: Slope of Incoming Line $= \text{Tan}[\theta_1]$

The slope of the tangent line to the curve $y = f[x]$ is the tangent of the angle it makes with the horizontal,

$$\text{Tan}[\theta_2] = f'[x]$$

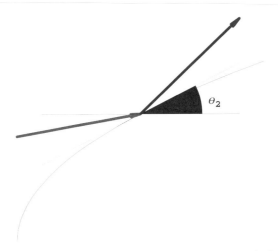

FIGURE 27.9: Slope of Tangent Line $= \text{Tan}[\theta_2]$

The angle between the first ray and the tangent line is $\theta_2 - \theta_1$.

HINT 27.10. *Let θ_3 denote the angle from the horizontal to the segment from $(x, f[x])$ to (u, v). Note the angles θ_1, θ_2, and θ_3 on the figure of a reflection off a curve.*

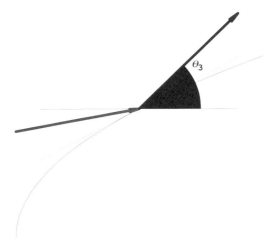

FIGURE 27.11: Slope of Reflected Line $= \text{Tan}[\theta_3]$

Show that Snell's law requires that $\beta = \theta_3 - \theta_2$, $\alpha = \theta_2 - \theta_1$, so

$$\theta_3 = 2\theta_2 - \theta_1$$

We can easily compute the tangents of the first two angles, since we can find the slopes.

HINT 27.12. *The trig identities*

$$\text{Tan}[\alpha \pm \beta] = \frac{\text{Tan}[\alpha] \pm \text{Tan}[\beta]}{1 \mp \text{Tan}[\alpha]\,\text{Tan}[\beta]}$$

allow us to find the slope of the reflection ray without actually finding any angles. Use the addition formulas for sine and cosine to prove these formulas, for example,

$$\text{Tan}[\alpha + \beta] = \frac{\text{Sin}[\alpha + \beta]}{\text{Cos}[\alpha + \beta]}$$
$$= \frac{\text{Sin}[\alpha]\,\text{Cos}[\beta] + \text{Sin}[\beta]\,\text{Cos}[\alpha]}{\text{Cos}[\alpha]\,\text{Cos}[\beta] - \text{Sin}[\alpha]\,\text{Sin}[\beta]}$$

Divide numerator and denominator by $\text{Cos}[\alpha]\,\text{Cos}[\beta]$ and simplify to prove the first formula for tangent.

Now use the tangent identity to find the slope of the reflected ray without actually computing angles.

HINT 27.13. *Write a short numerical computer program to find* $\text{Tan}[\theta_3]$ *if you assign a, b, x and have a defined function $f[x]$.*

$$t_1 = \frac{f[x] - b}{x - a} \quad (= \text{Tan}[\theta_1])$$
$$t_2 = f'[x] \quad (= \text{Tan}[\theta_2])$$
$$t_{22} = \frac{2\,t_2}{1 - t_2^2} \quad (= \text{Tan}[2\,\theta_2])$$
$$t_3 = ? \quad (= \frac{\text{Tan}[2\,\theta_2] - \text{Tan}[\theta_1]}{1 + \text{Tan}[2\,\theta_2]\,\text{Tan}[\theta_1]})$$

Test your program on the function $f[x] = 2\sqrt{x}$, when $a = 0$, $b = 6$, and $x = 4$. You should get:

$$-\frac{1}{2} = \text{Tan}[\theta_1]$$
$$\frac{1}{2} = \text{Tan}[\theta_2]$$
$$\frac{4}{3} = \text{Tan}[2\,\theta_2]$$
$$\frac{11}{2} = \text{Tan}[\theta_3]$$

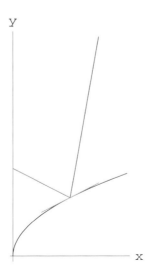

FIGURE 27.14: *Test Reflection*

Once we know the slope of the reflected ray, $t_3 = \text{Tan}[\theta_3]$, we can write an equation for that line. The formula "$y = m\,x + b$" or point-slope formula is the idea, but since we have used x at the point $(x, f[x])$ and since we know that the point $(x, f[x])$ lies on the line, we need new variables.

HINT 27.15. *Let dx represent a displacement away from the fixed point x so that $u = x + dx$. Show that the y-coordinate of the point (u, v) on the reflected ray is*

$$v = f[x] + t_3\, dx$$

We can use this in a computer program to draw the reflected ray from $(x, f[x])$ to (u, v). Test your program on the function $f[x] = 2\sqrt{x}$, when $a = 0$, $b = 6$, $x = 4$, and $dx = 2$. You should get $u = 6$ and $v = 15$.

Vertical Incoming Light

If the incoming light ray is vertical, $\theta_1 = \pi/2$, then $\text{Tan}[\theta_1]$ is undefined. The ray is perfectly OK, but the math breaks down.

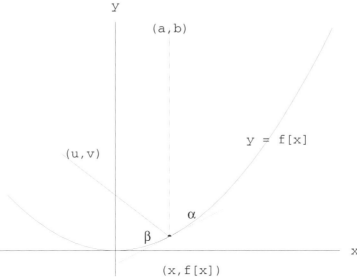

FIGURE 27.16: Vertical Incoming Light

HINT 27.17.

(1) Show that $\theta_3 = 2\theta_2 - \pi/2$ for vertical incoming light reflecting off the curve $y = f[x]$ whose tangent line makes the angle θ_2 with the vertical.
(2) Use the addition formulas for sine and cosine to show that $\text{Tan}[\pi/2 - \phi] = \frac{1}{\text{Tan}[\phi]}$.
(3) Show that
$$\text{Tan}[\theta_3] = \frac{(f'[x])^2 - 1}{2\, f'[x]}$$

EXPLORATION 27.18. *Use the formulas you obtained above to make computer programs that reflect a whole table of lines. For example, you might input a list of vertical lines beginning at $(x, 10)$ and hitting the graph $y = f[x]$ at $(x, f[x])$ for a list of x's. Each line in your input list generates a reflected line from your formula. Make reflections for several functions.*

Figure 27.19 is a comparison of rays of light reflected off different mirrors.

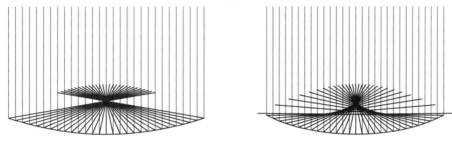

FIGURE 27.19: Parabolic Mirror   and   Spherical Mirror

You may have heard that a parabola focuses parallel light to a point.

EXPLORATION 27.20. *Prove that vertical rays reflected off a parabola all intersect at the "focus" point of the parabola.*

A sphere does not focus vertical rays to a point; it only focuses to a dark spot known as a caustic. What does an ellipse do? A sinusoid?

EXPLORATION 27.21. *A student discovered in his project that a sine curve nearly reflects light to a point. Can you think of a reason by using Taylor's formula from the Mathematical Background?*

Chapter 9

Applications in Economics

PROJECT 28

Monopoly Pricing

When there is only one firm that produces a certain product, that one firm is called a monopoly. Monopolies are familiar to most consumers. Local phone service and electric power are often provided by monopolies. Monopolists have the advantage over firms that must compete since, without regulation, they might have the ability to control the price of their service by controlling the quantity produced. A product that is in short supply will fetch a high price if people are demanding that product. Conversely, if the product is easy to come by, the price of it will be low. Monopolies could be detrimental to the consumer if they were interested not in providing enough of their product for everyone but in providing just enough to maximize their profits.

Some local monopolies that have received a lot of attention lately are cable TV franchises. People have been dissatisfied with having to pay extra fees for special cable channels. The question of regulating the cable companies has been a matter of some concern. In this project, you will be able to examine how monopolies set prices and answer for yourself whether or not the cable companies should be regulated by the government.

28.1. Going into Business

Suppose a small town has offered to give you the rights to provide cable TV service to families in the town. As a merchandiser you are interested in maximizing your profits. You are told that there are 100 families in the town who do not have cable TV. The cost to you of providing cable TV is $20 per month per family as well as $2000 in monthly overhead that is related to maintenance of your equipment and does not depend on how many families you service.

Fifty families live in houses and fifty families live in apartments. It has been estimated that people living in houses are more desirous of having cable TV. If you charge a price of p for a cable TV hookup, the following expression gives the number of families living in houses, q_h, who will pay for a hookup.

$$q_h = \begin{cases} 50, & \text{if } p \leq 100 \\ \frac{200-p}{2}, & \text{if } 100 \leq p \leq 200 \\ 0, & \text{if } p \geq 200 \end{cases}$$

The following expression for q_a gives the number of families living in apartments that will pay for a cable TV hookup if you charge a price p.

$$q_a = \begin{cases} \frac{150-p}{3}, & \text{if } p \leq 150 \\ 0, & \text{if } p \geq 150 \end{cases}$$

Economists usually call these demand curves (or demand functions). The town that is giving you the franchise will only allow you to set one price for cable TV. These "piecewise defined" functions may be defined on the computer, and there is help in the program **MonopolyHelp** on our website.

HINT 28.1. *Enter the computer commands in **MonopolyHelp** to make a plot of the demands by households and apartments as in Figure 28.2.*

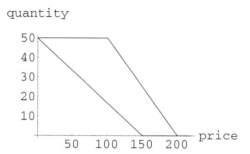

FIGURE 28.2: Demands for Cable TV

Your first task as monopolist is to determine what price you would charge for cable TV. We will determine this by solving a maximization problem in the exercises below. If you decide to charge a price p for cable TV, how many families in houses will buy a cable TV hookup? How many families living in apartments will buy a hookup? Determine expressions in terms of the price, p, of cable TV. These expressions will depend on what interval the price p lies in.

HINT 28.3. *Your revenue is the amount of money that you take in from cable TV hookups. It is equal to the price you charge times the number of cable TV hookups you sell. Express the revenue from sales to houses in terms of the price, p, you charge for hookups. What is the revenue from sales to apartments? Plot revenue as a function of price as in Figure 28.4.*

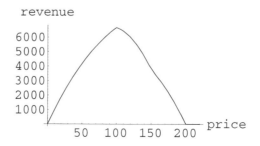

FIGURE 28.4: Cable Revenue

HINT 28.5. *You next must consider the costs incurred in providing cable service. The costs are $20 per hookup per month plus $2000 in maintenance costs. If you charge a price, p, for cable TV, what are the costs to you in providing cable TV to every family that demands it? Express your costs in terms of the price of cable TV, p. Plot total costs as a function of the unit price charged (Figure 28.6).*

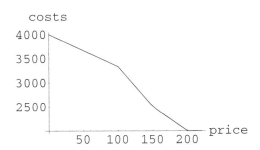

FIGURE 28.6: Cost for Providing Cable TV

HINT 28.7. *You are now ready to go into business (or are you?). The profit from running your cable company is your total revenue from sales minus the cost to you of providing the service. Express you profit as a function of the price, p, you charge for a cable TV hookup. Plot this. You wish to find the price to charge that will maximize your profit.*

Since your profit function is defined in pieces, you must solve more than one maximization problem. Your profit is a continuous function of your price, so the Extreme Value Theorem guarantees that there is a price that will maximize you profits. Why is it continuous? What are the endpoints for you maximization problems?

HINT 28.8. *Solve each maximization problem and determine the price you would charge to maximize your profits. How does the Extreme Value Theorem guarantee that you have found the point of maximum profits? You may wish to include a graph of your profit function to display the point of maximum profits (Figure 28.9).*

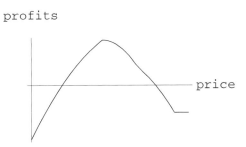

FIGURE 28.9: Profit for Providing Cable TV at a Single Price

HINT 28.10. *At the price you are going to charge for cable TV, how many families in homes are paying for hookups? How many families in apartments are paying for cable TV? Suppose the town, instead of giving you the right to provide cable TV, demands that you pay a monthly fee. What is the largest such fee that you will agree to pay?*

28.2. Going into Politics

Suppose instead of going into business, you are part of the government of the small town that has given someone the right to provide cable TV. The cable TV company has been in business several years and has led to some complaints. The biggest complaint has come from people living in apartments. They complain that the price of a cable TV hookup is so high that they can't afford it, and they want you to do something that will allow them to enjoy cable TV as well. The monopolist claims that he must charge a high price to remain in business but suggests that you allow him to charge different prices to apartments and homes.

HINT 28.11. *Suppose the monopolist can charge a price p_a to families living in apartments and a price p_h to families living in homes.*

What pair of prices maximizes profit? How many houses and apartments are served at these prices?

How large a service fee could the town charge and still allow you to stay in business?

PROJECT 29

Discrete Dynamics of Price Adjustment

This project studies the model economy mentioned at the beginning of Chapter CD20, Subsection 20.1.1, of the text, where there is a discrete price adjustment mechanism. Your project should begin with a description of that economy written in your own words and then contain a review of your baisc work on it from Chapter 20. We will set the model up again here and help you organize your review, but add an important economic term, the unit cost of production.

29.1. The Story

A new breakfast roll called 'Byties' becomes popular in our community. There is no patent on Byties, so anyone who chooses may decide to bake them. (Unlike a monopoly.) The price at which Byties sell determines who bakes Byties and how many they bake on a given day. At a high price, more people are willing to get up at 3:00 am, make the batter, and bake the rolls, so supply is an increasing function of price. For simplicity, we begin with the idealized assumption that the aggregate supply function is linear. "Aggregate supply" means that we don't know exactly who bakes the Byties but rather only know that the community's supply quantity, q, is given by

$$q = S[p] = a_S(p - c)$$

where c is the unit cost of ingredients, baking energy, etc., and a_S is a positive constant determined by the aggregate of bakers. For example, if the cost of producing one Bytie is 40 cents and $a_S = 1000$, then $S[p] = 1000(p - .40) = 1000\,p - 400$ (for p in dollars) as in the first example of Chapter 20. It is still useful to write supply in the form $1000(p - .40)$ because it tells us the cutoff point for bakers. No one will bake at a loss, so if the price drops below the cost of production, supply is zero. In other words, $q = S[p] = a_S(p - c)$ for $p \geq c$ and $S[p] = 0$ otherwise.

Similarly, our mythical community has a linear aggregate demand function. The total number of Byties purchased on a given day is given by

$$q = D[p] = b_D - a_D \cdot p$$

for positive constants a_D and b_D. Again, there are implicit constraints. No one sells at negative prices, so the function is only economically defined for $p \geq 0$ and, in practice, really only for $p \geq c$. Similarly, negative demand does not make sense, so there is an upper limit on the price our market will bear.

Basic discussion and understanding of your economy can begin with the specific case studied in Chapter 20

$$S[p] = 1000(p - .40)$$
$$D[p] = 1000 - 500p$$

but should strive to understand the role of the general parameters.

HINT 29.1. *The Model*
Describe the basic model, give the variables with units, and describe the role of the various parameters (a_S, c, a_D, a_S). The unit cost c is easy to explain, but you can at least say what large values of a_S mean about your economy. Other meaning may be derived indirectly. For example, what is the maximum price the market will bear in terms of the parameters you have chosen?

Does the demand function represent the wishes of every individual separately (multiplied by the number of people) or only the overall impact? For example, does the model say whether everyone buys twice as many Byties at 50 cents as at 1 dollar, or that more people buy at the lower price?

When supply equals demand, all the rolls are sold for producers willing to bake at that price and all the consumers willing to pay the price are satisfied. This is an equilibrium between the aggregates of producers and consumers.

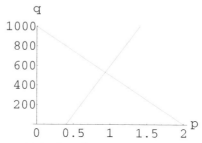

FIGURE 29.2: Supply and Demand for Byties

HINT 29.3. *Equilibrium Price*
What is the equilibrium price in terms of the parameters of your model? $p_e = ?$ The formula is easy and you will probably want to include this fundamental computation in your computer program so you can see the specific output for each choice of parameters. You could be lazy and ask the computer to solve the equations.

We have given you a start at the computations in the computer program **BytieHelp** on our website.

When demand exceeds supply, such as when all the rolls are sold well before breakfast time is over, there is incentive for more bakers to bake and charge a higher price. Our model economy adjusts price by making the change in price proportional to the amount by which demand exceeds supply. Suppose today's price is $p[t]$ and the excess demand $D[p[t]] - S[p[t]]$ is positive. The change in price for tomorrow, $p[t+1] - p[t]$, is a constant times the excess demand, so

$$p[t+1] = p[t] + k(D[p[t]] - S[p[t]])$$

where k is a positive constant.

HINT 29.4. *The Price Sequence*
What does the model say happens if supply exceeds demand?
Explain how the model predicts a sequence of daily prices, once you are given a starting price, that is, how an initial price $p[0] = P_0$ together with the change equation above determines a whole infinite sequence of prices $\{p[0], p[1], p[2], p[3], \cdots\}$. Does this basic determination require that supply and demand are linear? When they are linear, is there a closed formula for the value of the price on day t?

What is the importance of the parameter k for the stability of the breakfast economy of your mythical community? Use the computer program **BytieHelp** to compute and graph some examples where k is small and prices above equilibrium adjust downward toward a limit

$$\lim_{t \to \infty} p[t] = p_e$$

or prices below equilibrium adjust upward toward the equilibrium price. Give larger examples of k where prices oscillate above and below equilibrium, but still tend to equilibrium in the long term. Finally, give values of k where prices do not tend toward equilibrium. Figure 29.5 shows an example with $k = 0.0013$ showing both the cobweb and the explicit graph.

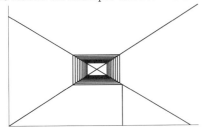
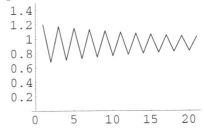

HINT 29.5. *Stability*
Use the Linear Stability Theorem of Chapter 20 to give a condition on your parameters to guarantee that every initial price results in a price sequence satisfying

$$\lim_{t \to \infty} p[t] = p_e$$

What is the best combined measure of stability for your economy? How do the parameters k, a_S, a_D, etc. affect the stability of the economy? Large values of a_S mean that producers are very responsive (why?), but does that make the economy more or less stable? Large values of a_D mean that consumers are very price sensitive (why?), but does that make the economy more or less stable? What combination of these sensitivities makes for a stable price sequence? Does the unit cost of production c affect the stability? Does it affect the limiting price?

29.2. The Basic Linear Model

Your project should summarize the results of the exercises so far for linear supply and demand. If you think about it, the exercises to here have told you how your economy behaves and the whole thing can be summarized quite briefly once you understand it all. This is a major portion of this project. Mathematical formulas can compress a lot of meaning into a small space. The remainder of the project explores a cost increase that we think of as taxation.

29.3. Taxation in the Linear Economy (Optional)

Your school wants to keep tuition down but must pay its bills and decides to put a tax on the popular new Byties. This increases the cost of production from c_1 to c_2. For example, in the numerical case at the beginning of the project you might consider the change from $S[p] = 1000(p - .40)$ to $S[p] = 1000(p - .45)$.

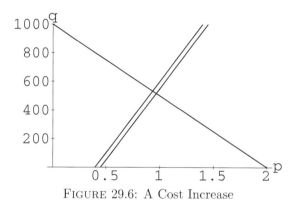

FIGURE 29.6: A Cost Increase

A natural question, even in this specific case, is, What happens to profits? The total profit before is the unit profit $p - c_1$ times the number sold at price p. The number sold is the demand, so total daily profit is ???

HINT 29.7. *Profit Experiments*
Add a computation to your computer program to compute the sequence of daily profits. What happens to profit for a 5 cent tax on Byties in the basic numerical case?

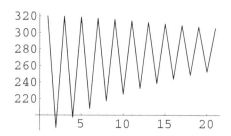

FIGURE 29.8: Daily Profits at $c = .40$

HINT 29.9. *Price Increases*
When unit cost is increased from c_1 to c_2, what is the resulting change in equilibrium price, p_{e1} to p_{e2}?

What is the effect of the change in c on the quantity sold at equilibrium? (Hint: Show that $q_{e1} = a_S(b_D - a_D \cdot c_1)/(a_S + a_D)$.)

Can a change in price affect the stability of the economy? For example, could the economy stably approach the first equilibrium price, but become unstable after a change from c_1 to c_2?

What is the effect of the change in c on profit at equilibrium prices? You can express the equilibrium profit in terms of the parameters and consider it a function of c, Profit[c] = ? What are the maximum and minimum of this function?

The previous exercise shows that the tax increases equilibrium price, decreases equilibrium demand, and decreases profit. The next question is, Who pays? The tax collected at equilibrium is $(c_2 - c_1)D[p_{e2}]$ and the change in profit is Profit[c_1] - Profit[c_2].

HINT 29.10. *Who Pays?*
Compare the loss of profit with the total tax collected.

29.4. A Nonlinear Economy (Optional)

Suppose that Byties are so delicious that, at least at low prices, folks don't care so much what they cost. We would have a demand function that was flatter at lower prices. A simple supply and demand pair with this feature is

$$q = S[p] = a_S(p - c)$$
$$q = D[p] = b_D - a_D p^2$$

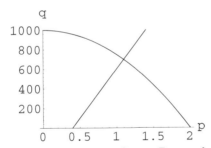

FIGURE 29.11: Nonlinear Demand

This is comparable to our first numerical example above with the following parameter values

$$q = S[p] = 1000(p - .40)$$
$$q = D[p] = 1000 - 250 p^2$$

We want you to study the effect of flattened nonlinear demand. Begin by answering the same questions you solved in the linear case.

HINT 29.12. *Nonlinear Basics*

(1) *What is the meaning of the model?* Here it would be sufficient to describe how the nonlinear model differs from your linear description. For example, the highest price the market will bear is the place where $D(p_h) = 0$. We used $a_D = 250$ to keep $p_h = \$2.00$ as in our first linear numerical case.

(2) *What is the general equilibrium price?* You can either solve algebraically or with Mathematica.

(3) The dynamics are still given by

$$p[0] = P_0$$
$$p[t+1] = p[t] + k(D[p[t]] - S[p[t]])$$

How does this determine the whole price sequence? Is there a closed form expression for the sequence?

(4) Use the Nonlinear Stability Theorem from Chapter 20 to give a criterion for stable price adjustment in the nonlinear economy. Run some experiments with

$$a_S = 1000 \qquad c = .40$$
$$a_D = 225 \qquad b_D = 1000$$

and various values of k to illustrate stability and instability. Repeat your experiments with $a_D = 245$ and $a_D = 275$ and test your stability criterion.

(5) Include a computation of daily profits in your stability experiments.

Here is a sample experiment showing price adjustments in both the cobweb and as an explicit graph.

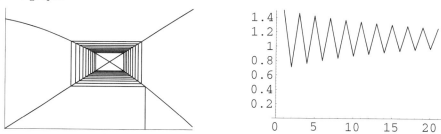

The initial price was \$1.50 and the factor for rate of adjustment was $k = 0.0013$.

29.5. Taxation in the Nonlinear Economy

Finally, we come to the questions concerning the addition of a tax on Byties. Begin your work with the following two experiments.

HINT 29.13. *Nonlinear Stability Experiments*
Let $k = 0.0013$, $a_S = 1000$, $a_D = 250$, $b_D = 1000$ and compute the equilibrium price and some price sequences for the two cases:
$c = .30$
$c = .50$
You will observe stability at 30 cents and instability at 50 cents. Why is this (mathematically and economically)?

What is the equilibrium profit at a unit cost of 30 cents?
What is the total tax collected if 20 cents is added to the 30 cent cost?

What is the profit at a unit cost of 50 cents?

How does the drop in profit from 30 to 50 cent costs compare with the total tax collected?

Conclude your project on the Bytie economy with some general comments about the difference between linear and nonlinear models.

PROJECT 30

Continuous Production and Exchange

This model examines what happens in an economy with production and exchange. In order to do this, we model a society with two producers, a woman and a man. Each produces one good and trades a portion of his/her goods for the other's goods. One key idea in the model is a simple 'law of diminishing returns' from consumption. Another is the balance between the satisfaction of consumption and the dissatisfaction of work. The third is the relative pleasure one gets from one's own goods.

VARIABLES

x	=	Goods produced by the woman in units of labor required
y	=	Goods produced by the man in units of labor required
$W[x,y]$	=	The woman's total utility from consumption of x and y
$M[x,y]$	=	The man's total utility from consumption of x and y

PARAMETERS

p	=	fraction each keeps of his/her own goods		
q	=	$1-p$	=	fraction each trades of his/her own goods
r	=	constant of proportionality associated with the dissatisfaction of work		

30.1. Why Trade?

Why don't producers simply produce and consume their own goods? By looking at the economy we live in today that should be somewhat obvious. Many of us would be very hungry and cold if we were forced to grow or catch our own food (even cook our own food from scratch), make our own clothes, and build our own house. The question is: How much to trade? and How does this decision affect the economy?

Although it is not necessary to assume that one unit of x is traded for one unit of y, we will assume that producers trade equal portions of work. This is a fairly reasonable assumption. Say they each spend 8 hours a day producing their goods. Would X be likely to give Y 4 units of x that took 4 hours each to produce in exchange for 4 units of y that took Y only 1 hour to produce? Probably not. They would be most likely to trade equal proportions of work. Therefore, we will assume that they each trade a fraction q of their work, keeping $p = 1 - q$.

A "utility function" gives a mathematical way to explain why people would exchange goods. Utility can be thought of as satisfaction. To the woman, it is more satisfying to have 75 units of her product and 25 units of the man's product than it is to just have 100 units of her own product. The 25 units of y have greater utility for her than the additional units of x, $W[75, 25] > W[100, 0]$.

In modeling production and exchange, it will be necessary to assume that utility can be quantified. This is not a very realistic assumption. Who is to say how many "units of utility" a chocolate bar is worth? Or for that matter, what a "unit of utility" is? However, we do have some sort of feel for how much utility a certain item or action gives us. For example, say someone were to offer you a new car. Wouldn't it be fair to say that you would receive more utility from a 1994 Mercedes convertible that you would receive from a 1981 Yugo? How much more? Maybe double the utility, maybe 5 or 10 times the utility. (It would depend on each individual.) So let's say you'd receive 5 times the utility from the Mercedes. If we arbitrarily assigned the Yugo 10 units of utility, then we would say the Mercedes is worth 50 units of utility. The property we have just described says that some inputs to the utility function are worth more than others in terms of output. We need our model utility function to have this property.

The next assumption involves diminishing utility. This states that each successive unit adds less to the total utility than the previous unit. For example, that Mercedes may have given you 50 units of utility the first time. However, once you already have five Mercedes, is another one just like it going to give you an additional 50 units of utility? Probably not. If this doesn't seem quite so obvious, think of someone offering you another ice cream cone after you have just eaten five ice cream cones. Would that sixth cone give you as much utility as the first one?

If ownership is satisfying, but by a decreasing amount as goods increase, how can we quantify it? "Fechner's law" (that we saw in the Richter Scale Problem 21.1 of the text) can help us with that. This says that the more you have, the less satisfied you are with new goods. If a change in utility U is the differential dU and a change in goods g is dg, then this law says that the change in satisfaction, dU, for a fixed size change in goods, dg, gets smaller as g gets larger.

HINT 30.1. *Logarithmic Returns*
1) Explain in what sense the equation

$$dU = k\frac{dg}{g}$$

means that, "the more you have, the less you are satisfied with new goods." How satisfied are you with the first tiny bit of goods when you don't own any?
2) Integrate both sides of the above expression to find an expression for utility

$$U = ?$$

How satisfied are you if you don't own any goods?
3) We need to correct the mathematical problem with our first attempt at a law of diminishing satisfaction. Suppose we can subsist on 1 unit of goods. Positive satisfaction arises from

increased goods. Explain why the law

$$dU = k\frac{dg}{1+g}$$

means that, "the more you have, the less you are satisfied with new goods, beyond subsistence."

4) Integrate both sides of this law and use the fact that $U = 0$ if $g = 0$ to show

$$U = k\operatorname{Log}[1+g]$$

Another assumption, at least in the first part of the project, is that work is unsatisfying. If this is the case, work contributes a negative satisfaction. The simplest model of dissatisfaction is that it is proportional to the work done. Let $W[x,y]$ and $M[x,y]$ denote the satisfaction or utility from production and exchange for the woman and man, respectively. From the assumptions above, we can see that when we assume work is unsatisfying, work would contribute $-rx$ and $-ry$ to W and M, respectively, where r is a constant.

Our law of diminishing satisfaction will tell us the other component of utility. The woman subsists, keeps px of her goods, and exchanges qy of the man's goods. Her total goods then are $1 + px + qy$.

HINT 30.2. *Describe the meaning of the utility functions*

$$W[x,y] = \operatorname{Log}[1+px+qy] - rx$$
$$M[x,y] = \operatorname{Log}[1+qx+py] - ry$$

for the man and the woman. Do they receive equal satisfaction from a given level of production?

Now we have an model for satisfaction, or utility, but what does that tell us? Using the equations for satisfaction, along with the assumption that neither X nor Y will increase production unless his/her satisfaction is increased, we can formulate equations for $\frac{dx}{dt}$ and $\frac{dy}{dt}$.

HINT 30.3. *Explain the Model*
Explain the dynamics of the economic model

$$\frac{dx}{dt} = c\left(\frac{p}{px+qy+1} - r\right) = c\frac{\partial W}{\partial x}$$
$$\frac{dy}{dt} = c\left(\frac{p}{qx+py+1} - r\right) = c\frac{\partial M}{\partial y}$$

(We can choose a unit of time such that $c = 1$.)

Each producer can only produce her/his own good, so each considers only the effect that a change in her/his good would have on her/his satisfaction. How is this expressed mathematically in the dynamical system? Why do these independent decisions still effect both producers?

We want to explore the economic meaning of the relative sizes of the parameters p and r (recall that $q = 1 - p$.)

Now that you understand the way this economy adjusts production in order to improve itself, we want to investigate the dynamics. We would hope that each producer strives to make things better and that in time the economy approaches an ideal state, maximizing pleasure and minimizing work. There is some danger. The independent decisions might make one producer worse and worse off...

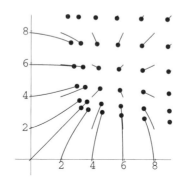

FIGURE 30.4: Dynamic Production

HINT 30.5. *Experiments*
The first step is to see how the model acts in general. These equations are autonomous. Why? This means that we can modify the computer program **Flow2D** to try some experiments on the economy. Begin with $p = 7/8$ and $r = 1/7$. This means two things. First, each producer likes his/her own good. Why? Second, neither is very work averse. Why? Try the case $p = 1/3$ and $r = 1/7$. What does this mean? What happens dynamically? Try the case $p = 7/8$ and $r = 1$. What does this mean? What happens dynamically?

Pleasing economic dynamics would be for the independent production adjustments to result in the whole economy moving toward a positive equilibrium. Negative production or some sort of borrowing is not really built into our basic assumptions, so we need to give conditions that mean our model makes sense.

HINT 30.6. *Equilibrium*
Show that there is a positive equilibrium when $p > r$. How could you describe this inequality in terms of economic preferences? (Hint: Use the computer to show that the equilibrium is $x = y = (p - r)/r$.)

Once you have found an equilibrium point, you can use compass heading arrows from hand sketched direction fields to determine the flow and stability of an equilibrium point. The computer can help you confirm the stability or instability using numerical experiments, but it can also perform the symbolic computations needed to characterize stability as follows.

HINT 30.7. *Stability*
Our production and exchange economy has a positive equilibrium if people would rather work and use their goods, $p > r$, than goof off and have nothing. This equilibrium is a stable attractor if each producer likes her/his own goods better than those of the other producer. What is the associated mathematical condition and why does it say this? What happens when it fails?

Chapter 10

Advanced Max-Min Problems

PROJECT 31

Geometric Optimization Projects

The distance between lines, between curves, between a curve and a surface, etc. is a multidimensional minimization problem. "The distance" means the shortest distance as the points vary over both objects. Most calculus books do not treat this topic, because the solution of the equations resulting in the critical point step is too hard to do by hand. The computer can help you solve these deep geometrical problems.

The linear cases of these distance problems can usually be solved with vector geometry and justified by the Pythagorean theorem, but it is worthwhile seeing the calculus solutions even in these cases.

31.1. Distance between Lines

Two lines are given by parametric equations as follows:

Line 1

$\mathbf{X}[s]:$
$x_1 = s$
$x_2 = 0$
$x_3 = 0$

Line 2

$\mathbf{Y}[t]:$
$y_1 = 3 + t$
$y_2 = 3 + t$
$y_3 = 1 + 2t$

Find the distance between these lines; that is, minimize the distance between a point on line 1 and a point on line 2. Analytically, we have the question,

FIND: The minimum of $|\mathbf{Y}[s] - \mathbf{X}[t]|$ for all values of s and t.

Since the square is an increasing function, we may as well solve the easier problem of minimizing the square of the distance

$$D[s,t] = |\mathbf{X}[s] - \mathbf{Y}[t]|^2 = (\mathbf{X}[s] - \mathbf{Y}[t]) \bullet (\mathbf{X}[s] - \mathbf{Y}[t])$$
$$= (y_1[t] - x_1[s])^2 + (y_2[t] - x_2[s])^2 + (y_3[t] - x_3[s])^2$$
$$= (3 + t - s)^2 + (3 + t - 0)^2 + (1 + 2t - 0)^2$$

The partial derivatives of D are given by the chain rule

$$\frac{\partial D}{\partial s} = 2(3 + t - s)(-1) \quad \text{and} \quad \frac{\partial D}{\partial t} = 2(3 + t - s) + 2(3 + t) + 2(1 + 2t)(2)$$

The critical point condition is the system of equations

$$\frac{\partial D}{\partial s} = 0 \qquad \text{or} \qquad 2s - 2t = 6$$
$$\frac{\partial D}{\partial t} = 0 \qquad \qquad -2s + 12t = -16$$

with single solution $(s, t) = (2, -1)$. The point $\mathbf{X}[2] = (2, 0, 0)$ is nearest to the point $\mathbf{Y}[-1] = (2, 2, -1)$ at the distance $\sqrt{0^2 + 2^2 + (-1)^2} = \sqrt{5}$.

Notice that the critical point condition for lines is a system of two linear equations in s and t. Solving this is a high school algebra problem.

HINT 31.1. *Why is the only critical point a minimum?*

You could use geometric or analytical reasoning to answer the last question. Analytically, we could introduce a "fake" compact set for the (s, t) domain, say $|s| \leq S$ and $|t| \leq T$ with both S and T huge. How large is $D[s, t]$ on the boundary of the fake domain? Why does $D[s, t]$ have both a max and min on the fake domain? Why is the min inside?

EXAMPLE 31.2. *Geometry of the Critical Point*

The vector $\mathbf{V}$ pointing from $\mathbf{X}[2] = (2, 0, 0)$ to $\mathbf{Y}[-1] = (2, 2, -1)$ is perpendicular to both lines,

$$\begin{aligned} \mathbf{V} &= \mathbf{Y}[-1] - \mathbf{X}[2] \\ &= (2, 2, -1) - (2, 0, 0) \\ &= (0, 2, -1) \end{aligned}$$

The direction of line 1 is $\mathbf{V}_1 = (1, 0, 0)$ and the direction of line 2 is $\mathbf{V}_2 = (1, 1, 2)$

$$\mathbf{V} \bullet \mathbf{V}_1 = (0, 2, -1) \bullet (1, 0, 0) = 0$$
$$\mathbf{V} \bullet \mathbf{V}_2 = (0, 2, -1) \bullet (1, 1, 2) = 0$$

Why must this perpendicularity be true?

HINT 31.3. *Critical Points*

(1) Find the distance between line 1 in the direction $(1, -1, 1)$ through the point $(0, 1, 0)$ and line 2 in the direction $(1, 2, 1)$ through the point $(2, 2, 2)$. Is the segment connecting the two minimizing points perpendicular to both lines?
(2) Find the distance between line 1 and line 3 in the direction $(1, 2, 1)$ through the point $(2, 1, 2)$. Is the segment connecting the two minimizing points perpendicular to both lines?
(3) Let line α be given by the parametric equations $\mathbf{X}[s] = \mathbf{P}_1 + s\mathbf{V}_1$ and line β be given by the parametric equations $\mathbf{Y}[t] = \mathbf{P}_2 + t\mathbf{V}_2$. We want to minimize

$$\begin{aligned} D[s, t] &= |(\mathbf{P}_1 + s\mathbf{V}_1) - (\mathbf{P}_2 + t\mathbf{V}_2)|^2 \\ &= (\mathbf{P}_1 - \mathbf{P}_2 + s\mathbf{V}_1 - t\mathbf{V}_2) \bullet (\mathbf{P}_1 - \mathbf{P}_2 + s\mathbf{V}_1 - t\mathbf{V}_2) \end{aligned}$$

Show that

$$\frac{\partial D}{\partial s} = 2\mathbf{V}_1 \bullet (\mathbf{P}_1 - \mathbf{P}_2 + s\mathbf{V}_1 - t\mathbf{V}_2)$$

$$\frac{\partial D}{\partial t} = -2\mathbf{V}_2 \bullet (\mathbf{P}_1 - \mathbf{P}_2 + s\mathbf{V}_1 - t\mathbf{V}_2)$$

Suppose that s_0 and t_0 are the critical solutions where these partials are zero. Show that the segment connecting $\mathbf{X}[s_0]$ and $\mathbf{Y}[t_0]$ is perpendicular to both $\mathbf{V}_1$ and $\mathbf{V}_2$,

$$(\mathbf{Y}[t_0] - \mathbf{X}[s_0]) \bullet \mathbf{V}_1 = 0$$
$$(\mathbf{Y}[t_0] - \mathbf{X}[s_0]) \bullet \mathbf{V}_2 = 0$$

(Hint: Substitute $\mathbf{X}[s_0] = \mathbf{P}_1 + s_0 \mathbf{V}_1$ and $\mathbf{Y}[t_0] = \mathbf{P}_2 + t_0 \mathbf{V}_2$ in the equations and compare with the partials.)

31.2. Distance between Curves

Find the distance between the ellipse $(x/3)^2 + (y/4)^2 = 1$ and the branch of the hyperbola $xy = 10$ with $x > 0$. Parametric equations for these curves are

Ellipse		Hyperbola	
$\mathbf{X}[s]$:	$x_1 = 3\operatorname{Cos}[s]$	$\mathbf{Y}[t]$:	$y_1 = t$
	$x_2 = 4\operatorname{Sin}[s]$		$y_2 = 10/t$

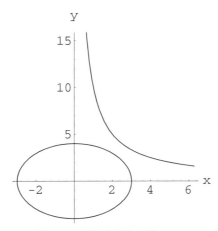

FIGURE 31.4: The Curves

FIND: The minimum distance between $\mathbf{X}[s]$ and $\mathbf{Y}[t]$, or the simpler square of the distance,

$$D[s,t] = |\mathbf{Y}[t] - \mathbf{X}[s]|^2 = [t - 3\operatorname{Cos}[s]]^2 + [10/t - 4\operatorname{Sin}[s]]^2$$

HINT 31.5. *Find the specific set of critical point equations for this minimization problem*

$$\frac{\partial D}{\partial s} = 0$$
$$\frac{\partial D}{\partial t} = 0$$

and solve them with the computer.

The computer's exact solution function may not be powerful enough to find exact solutions to the equations that arise from minimizing the distance between these curves, but the computer can approximate roots if you know a starting point. Look at Figure 31.4 above and say which roots are near the initial guess $s = 4$ and $t = 2$. Then calculate the result using a numerical solver such as *Mathematica's* **FindRoot[ ]** by making appropriate initial guesses. Remember that the computer can compute the partial derivatives for you as part of the solution program.

HINT 31.6. *We got solutions $(x_1, x_2) = (2.053, 2.917)$, $(y_1, y_2) = (2.827, 3.537)$ and $(x_1, x_2) = (-1.557, -3.419)$, $(y_1, y_2) = (3.515, 2.845)$. Are these the min and the max? Which pair of points minimize the distance between a point on the ellipse and a point on the branch of the hyperbola?*

Use the computer to help you be sure you have what you need. For example, Figure 31.7 is a plot of the distance as a function of the parameters.

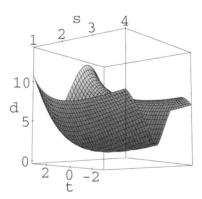

FIGURE 31.7: Distance versus (s, t)

HINT 31.8. *Is the segment connecting the minimizing points perpendicular to both curves?*

31.3. An Implicit-Parametric Approach

We can work with the implicit equation of the ellipse and the parametric equation of the hyperbola. First, we write the distance squared with a constraint,

$$D = (x-t)^2 + \left(y - \frac{10}{t}\right)^2 \quad : \quad \frac{x^2}{9} + \frac{y^2}{16} = 1$$

We want to treat D as a function of x and t where y is determined by the constraint.

$$dD = \frac{\partial D}{\partial x}\,dx + \frac{\partial D}{\partial y} + \frac{\partial D}{\partial t}\,dt \quad : \quad \frac{2x}{9}\,dx + \frac{2y}{16}\,dy = 0$$

$$: \quad dy = -\frac{16x}{9y}\,dx \quad (\text{provided } y \neq 0)$$

$$dD = \left(\frac{\partial D}{\partial x} - \frac{16x}{9y}\frac{\partial D}{\partial y}\right)dx + \frac{\partial D}{\partial t}\,dt$$

This differential is the local linear approximation to the change in the square distance function, so the critical values occur where it is the zero funciton.

HINT 31.9. *Solve the three equations*

$$\frac{\partial D}{\partial x} - \frac{16x}{9y}\frac{\partial D}{\partial y} = 0$$

$$\frac{\partial D}{\partial t} = 0$$

$$\frac{x^2}{9} + \frac{y^2}{16} = 1$$

with your computer's numerical equation solver. We got one solution $x = 2.053$, $y = 2.917$, $t = 2.827$.

Examples 19.9, 19.11, and the program **SvVarMxMn** from the main text find the distance from a point to a surface. We can generalize those examples to find the distance from a line or a curve to a surface.

31.4. Distance from a Curve to a Surface

We generalize the main text Example 19.9 to find the distance between the explicit surface $x_3 = x_1^2 + 2\,x_2^2$ and the parametric curve

$$y_1[t] = 3\,\text{Cos}[t]$$
$$y_2[t] = 4\,\text{Sin}[t]$$
$$y_3[t] = 1 - y_1[t] + y_2[t]$$

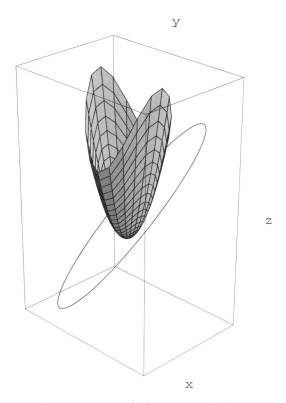

FIGURE 31.10: A Curve and Surface

The squared distance function is

$$D[x_1, x_2, t] = (x_1 - y_1[t])^2 + (x_2 - y_2[t])^2 + ((x_1^2 + 2\,x_2^2) - y_3[t])^2$$

HINT 31.11. *Numerically solve the equations*

$$\frac{\partial D}{\partial x_1} = 0$$
$$\frac{\partial D}{\partial x_2} = 0$$
$$\frac{\partial D}{\partial t} = 0$$

One solution we found was $x_1 = 0.979$, $x_2 = 0.744$, $t = 0.728$. Is this the minimum distance?

EXPLORATION 31.12. *You can easily invent any number of interesting distance problems yourself now and solve them with a combination of calculus and computing.*

PROJECT 32

Least Squares Fit and Max-Min

These questions have you complete the development of the least squares fit of data.

32.1. Introductory Example

The basic idea of least squares fit is this. We are given data points

$$(x_1, y_1), (x_2, y_2), \cdots, (x_n, y_n)$$

for example, $(1,6)$, $(2,9)$, and $(3,10)$. We know that these represent the output of a linear function

$$z = m\, x + b$$

with errors introduced by measurement. We do not know the parameters m and b but wish to estimate them so that they produce the minimal sum of squared errors. What are the squared errors?

Suppose we had chosen a value for m and a value for b. If we compute values using the formula $m\, x_i + b$ and subtract the measured value of y_i, we obtain the error at one point,

$$\begin{aligned}\text{Error at } (x_1, y_1) &= [\text{computed value} - \text{measured value}] \\ &= [(m\, x_1 + b) - y_1] \\ &= (m \cdot 1 + b) - 6 \quad \text{in the example}\end{aligned}$$

At the input $x_2 = 2$, we would compute output $z_2 = m \cdot 2 + b$ and compare that with $y_2 = 9$; the difference is $(m \cdot 2 + b) - 9$. In this example, the sum of all these actual errors is

$$[(m \cdot 1 + b) - 6] + [(m \cdot 2 + b) - 9] + [(m \cdot 3 + b) - 10]$$

This straight algebraic sum has the disadvantage that a huge positive error and a huge negative error cancel, so that the sum does not represent the "total error." (We do not want two huge canceling errors to be considered small.) The square of an error, like $[(m \cdot 2 + b) - 9]^2$, is always positive and in addition favors small errors,

$$\text{Squared error at } (x_1, y_1) = [(m\, x_1 + b) - y_1]^2$$

so the sum of the squared errors:

$$E = [(m \cdot 1 + b) - 6]^2 + [(m \cdot 2 + b) - 9]^2 + [(m \cdot 3 + b) - 10]^2$$

is a measure of overall error that does not cancel errors of opposite sign.

We want to consider the parameters m and b as the unknowns of our problem, so we seek to

$$\text{Minimize } [E[m, b] : \text{all real } m, b]$$

and specifically, we wish to minimize

$$E[m, b] = [(m \cdot 1 + b) - 6]^2 + [(m \cdot 2 + b) - 9]^2 + [(m \cdot 3 + b) - 10]^2$$

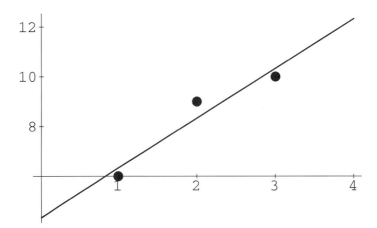

FIGURE 32.1: Least Squares Fit

HINT 32.2. *In this problem we want the minimum, but mathematically, we might also ask for the maximum. Does the function $E[m, b]$ have a maximum value over all real values of m and b? Why? Show that if either $m = \pm H$ or $b = \pm H$ is a huge positive or negative number, then E is huge.*

32.2. The Critical Point

The exercise above shows that $E[m, b]$ has no maximum. It does have a minimum, but before we examine the reason, we seek possible locations for the min, that is, critical values or places where the gradient vector is zero.

$$\frac{\partial E}{\partial m} = 2[(m \cdot 1 + b) - 6] + 2[(m \cdot 2 + b) - 9] \cdot 2 + 2[(m \cdot 3 + b) - 10] \cdot 3$$

$$\frac{\partial E}{\partial b} = 2[(m \cdot 1 + b) - 6] + 2[(m \cdot 2 + b) - 9] + 2[(m \cdot 3 + b) - 10]$$

so

$$\frac{\partial E}{\partial m} = 0 \Leftrightarrow (m \cdot 1 + b) \cdot 1 + (m \cdot 2 + b) \cdot 2 + (m \cdot 3 + b) \cdot 3 = 6 \cdot 1 + 9 \cdot 2 + 10 \cdot 3$$

$$\frac{\partial E}{\partial b} = 0 \Leftrightarrow (m \cdot 1 + b) + (m \cdot 2 + b) + (m \cdot 3 + b) = 6 + 9 + 10$$

or

$$(1 \cdot 1 + 2 \cdot 2 + 3 \cdot 3)\, m + (1 + 2 + 3)\, b = 6 \cdot 1 + 9 \cdot 2 + 10 \cdot 3$$
$$(1 + 2 + 3)\, m + (1 + 1 + 1)\, b = 6 + 9 + 10$$

This is a system of two linear equations in the unknowns m and b,

$$a \cdot m + d \cdot b = f$$
$$c \cdot m + e \cdot b = g$$

or in the more compact matrix notation $A \cdot X = B$,

$$\begin{bmatrix} a & d \\ c & e \end{bmatrix} \begin{bmatrix} m \\ b \end{bmatrix} = \begin{bmatrix} f \\ g \end{bmatrix}$$

where $a = (1^2 + 2^2 + 3^2) = 14$, $c = d = (1 + 2 + 3) = 6$, $e = 3$, $f = 1 \cdot 6 + 2 \cdot 9 + 3 \cdot 10 = 54$ and $g = 6 + 9 + 10 = 25$. The unknowns are m and b.

We have just shown that $\nabla E = 0$ if and only if $A \cdot X = B$, with the particular matrices,

$$\begin{bmatrix} 14 & 6 \\ 6 & 3 \end{bmatrix} \begin{bmatrix} m \\ b \end{bmatrix} = \begin{bmatrix} 54 \\ 25 \end{bmatrix}$$

HINT 32.3. *The unique solution of these particular linear equations is $(m, b) = (2, \frac{13}{3})$. Verify this by a hand computation.*

HINT 32.4. *Why is the solution of the critical point equations the parameter values of minimum total square error? In other words, how do you know that there is a minimum pair of values and not just better and better choices? Why is this point the minimum one? Use the Extreme Value Theorem and Critical Point Theorem, for example, to answer the question.*

32.3. The General Critical Equations

We want you to derive the general formulas for the minimizing m and b. Suppose that n pairs of (x, y) values are given,

$$(x_1, y_1), (x_2, y_2), \cdots, (x_n, y_n)$$

Let

$$E[m, b] = [(m \cdot x_1 + b) - y_1]^2 + [(m \cdot x_2) + b - y_2]^2 + \cdots + [(m \cdot x_n + b) - y_n]^2$$

HINT 32.5. *Show that $\nabla E = 0$ if and only if the unknowns (m, b) satisfy the equations $A \cdot X = B$,*

$$\begin{bmatrix} a & d \\ c & e \end{bmatrix} \begin{bmatrix} m \\ b \end{bmatrix} = \begin{bmatrix} f \\ g \end{bmatrix}$$

where $a = (x_1^2 + x_2^2 + x_3^2 + \cdots + x_n^2)$, $c = d = (x_1 + x_2 + x_3 + \cdots + x_n)$, $e = n$, $f = x_1 \cdot y_1 + x_2 \cdot y_2 + x_3 \cdot y_3 + \cdots + x_n \cdot y_n$, *and* $g = y_1 + y_2 + y_3 + \cdots + y_n$

Notice the n-dimensional norm and dot product in these formulas!

HINT 32.6. *Write a computer program to find the least squares fit of the data* $(1, 6)$, $(1.2, 7.1)$, $(1.3, 7.2)$, $(1.35, 7.2)$, $(1.41, 7.5)$, $(1.5, 8)$, $(1.53, 8.2)$, $(1.6, 8.9)$. *Use the program* **LeastSquares** *on our website for help.*

In the example with the data $(1, 6)$, $(2, 9)$, and $(3, 10)$ and least squares fit $z = 2\,x + \frac{13}{3}$, show that the sum of the signed errors is zero:

$$\left(2 \cdot 1 + \frac{13}{3} - 6\right) + \left(2 \cdot 2 + \frac{13}{3} - 9\right) + \left(2 \cdot 3 + \frac{13}{3} - 10\right) = 0$$

In other words, the amount above and below the fit line is the same.

HINT 32.7. *Show that the sum of the signed errors is zero in the general case.*

PROJECT 33

Local Max-Min and Stability of Equilibria

This chapter uses the local stability of a dynamical system to classify critical points as local maxima or minima.

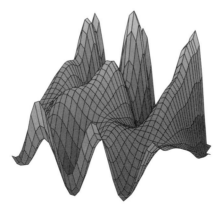

FIGURE 33.1: A Smooth Mountain Range

33.1. Steepest Ascent

An ambitious but nearsighted climber on a mountain range of height h,

$$z = h[x, y]$$

will move in the direction of fastest increase, ∇h. This is nearsighted in that he may head toward a local maximum or smaller peak, instead of crossing a pass or saddle and climbing the highest peak in the range.

The statement that "the climber moves in the direction of steepest ascent" can be expressed by the dynamical system
$$\frac{d\mathbf{X}}{dt} = \nabla h[\mathbf{X}]$$
or in components
$$\frac{dx}{dt} = \frac{\partial h}{\partial x}[x, y]$$
$$\frac{dy}{dt} = \frac{\partial h}{\partial y}[x, y]$$

A solution of these differential equations $(x[t], y[t])$ gives the path of the climber as a function of time by using $z[t] = h[x[t], y[t]]$.

These differential equations stop a path at a critical point (x_c, y_c) because the critical point condition
$$\frac{\partial h}{\partial x}[x_c, y_c] = 0$$
$$\frac{\partial h}{\partial y}[x_c, y_c] = 0$$
is the same as the equilibrium condition of $\frac{d\mathbf{X}}{dt} = \nabla h$.

33.2. The Second Derivative Test in Two Variables

We can use the local stability criteria Theorem 24.6 of the main text to test whether a critical point is a max or min. This is only a nearsighted local test, because the linearization of the dynamical system is only local. The linearization of $\frac{d\mathbf{X}}{dt} = \nabla h$ at (x_c, y_c) is

$$\begin{bmatrix} \frac{du}{dt} \\ \frac{dv}{dt} \end{bmatrix} = \begin{bmatrix} \frac{\partial f}{\partial x}[x_c, y_c] & \frac{\partial f}{\partial y}[x_c, y_c] \\ \frac{\partial g}{\partial x}[x_c, y_c] & \frac{\partial g}{\partial y}[x_c, y_c] \end{bmatrix} \begin{bmatrix} u \\ v \end{bmatrix}$$

where $f[x, y] = \frac{\partial h}{\partial x}$ and $g[x, y] = \frac{\partial h}{\partial y}$. This makes the linear system matrix

$$\begin{bmatrix} \frac{\partial^2 h}{\partial x^2}[x_c, y_c] & \frac{\partial^2 h}{\partial y \partial x}[x_c, y_c] \\ \frac{\partial^2 h}{\partial x \partial y}[x_c, y_c] & \frac{\partial^2 h}{\partial y^2}[x_c, y_c] \end{bmatrix}$$

Continuous second partial derivatives are always symmetric, $\frac{\partial^2 h}{\partial x \partial y} = \frac{\partial^2 h}{\partial y \partial x}$, so the characteristic equation of the linear dynamical system is

$$r^2 - \left(\frac{\partial^2 h}{\partial x^2}[x_c, y_c] + \frac{\partial^2 h}{\partial y^2}[x_c, y_c]\right) r + \left(\frac{\partial^2 h}{\partial x^2}[x_c, y_c] \frac{\partial^2 h}{\partial y^2}[x_c, y_c] - \left(\frac{\partial^2 h}{\partial x \partial y}[x_c, y_c]\right)^2\right) = 0$$

An equation
$$r^2 - br + c = 0$$
has roots of the same sign if
$$b + \sqrt{b^2 - 4c} \quad \text{and} \quad b - \sqrt{b^2 - 4c}$$
have the same sign.

HINT 33.2. *The Second Derivative Test*

(1) *Show that a critical point* (x_c, y_c) *of* $z = h[x, y]$ *is neither a local max nor a local min if*

$$\left(\frac{\partial^2 h}{\partial x^2}[x_c, y_c] \frac{\partial^2 h}{\partial y^2}[x_c, y_c] - \left(\frac{\partial^2 h}{\partial x \partial y}[x_c, y_c]\right)^2\right) < 0$$

(2) *Show that a critical point* (x_c, y_c) *of* $z = h[x, y]$ *is a locally attracting equilibrium for* $\frac{d\mathbf{X}}{dt} = \nabla h$ *if*

$$\left(\frac{\partial^2 h}{\partial x^2}[x_c, y_c] \frac{\partial^2 h}{\partial y^2}[x_c, y_c] - \left(\frac{\partial^2 h}{\partial x \partial y}[x_c, y_c]\right)^2\right) > 0$$

and

$$\frac{\partial^2 h}{\partial x^2}[x_c, y_c] + \frac{\partial^2 h}{\partial y^2}[x_c, y_c] < 0$$

Why is an attractor of the system $\frac{d\mathbf{X}}{dt} = \nabla h$ *a local maximum for the height function* $z = h[x, y]$?

(3) *Show that a critical point* (x_c, y_c) *of* $z = h[x, y]$ *is a locally repelling equilibrium for* $\frac{d\mathbf{X}}{dt} = \nabla h$ *if*

$$\left(\frac{\partial^2 h}{\partial x^2}[x_c, y_c] \frac{\partial^2 h}{\partial y^2}[x_c, y_c] - \left(\frac{\partial^2 h}{\partial x \partial y}[x_c, y_c]\right)^2\right) > 0$$

and

$$\frac{\partial^2 h}{\partial x^2}[x_c, y_c] + \frac{\partial^2 h}{\partial y^2}[x_c, y_c] > 0$$

Why is a repellor of the system $\frac{d\mathbf{X}}{dt} = \nabla h$ *a local minimum for the height function* $z = h[x, y]$?

(4) *Prove the algebraic identity*

$$(a + b)^2 - 4ab + c^2 = (a - b)^2 + c^2$$

and use it to show that the characteristic equation of this gradient dynamical system cannot have complex roots. Can you imagine a mountain that spirals you into a max or min by steepest ascent or descent?

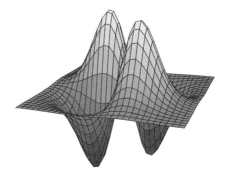

FIGURE 33.3: $z = x\,y\,e^{-\frac{1}{2}(x^2+y^2)}$

There are a lot of formulas, so let's review the steps in classifying local extrema.

(1) Compute the symbolic first partial derivatives

$$\frac{\partial h}{\partial x}$$
$$\frac{\partial h}{\partial y}$$

(2) Use the first partials to compute the second partials,

$$\begin{bmatrix} \frac{\partial^2 h}{\partial x^2} & \frac{\partial^2 h}{\partial y \partial x} \\ \frac{\partial^2 h}{\partial y \partial x} & \frac{\partial^2 h}{\partial y^2} \end{bmatrix}$$

(3) Solve the equations

$$\frac{\partial h}{\partial x} = 0$$
$$\frac{\partial h}{\partial y} = 0$$

for all critical points, (x_c, y_c).

(4) Substitute each of these solutions into the second partial derivatives

$$\begin{bmatrix} \frac{\partial^2 h}{\partial x^2} & \frac{\partial^2 h}{\partial y \partial x} \\ \frac{\partial^2 h}{\partial y \partial x} & \frac{\partial^2 h}{\partial y^2} \end{bmatrix} [x_c, y_c]$$

and apply the test in the previous exercise to these numbers.

HINT 33.4. *Local Max-Min*
Let $z = h[x,y] = x\,y\,e^{-\frac{1}{2}(x^2+y^2)}$, so $\frac{\partial h}{\partial x} = 0$ and $\frac{\partial h}{\partial y} = 0$ at $(x_c, y_c) = (0,0), (1,1), (1,-1), (-1,1),$ and $(-1,-1)$. Show that the matrix of second partial derivatives is

$$e^{-\frac{1}{2}(x^2+y^2)} \begin{bmatrix} x\,y\,(x^2-3) & x^2\,y^2 - x^2 - y^2 + 1 \\ x^2y^2 - x^2 - y^2 + 1 & x\,y\,(y^2-3) \end{bmatrix}$$

and use the max-min criteria to classify these critical points.

Classify the critical points of $z = 3(x^2 + 3\,y^2)\,e^{-(x^2+y^2)}$.

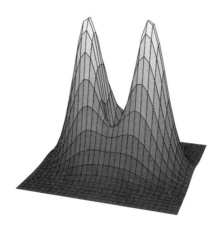

FIGURE 33.5: $z = 3(x^2 + 3\,y^2)\,e^{-(x^2+y^2)}$

Chapter 11

Applications of Linear Differential Equations

The following projects use linear differential equations.

PROJECT 34

Lanchester's Combat Models

Frederick William Lanchester was an English engineer who died after the end of World War II in 1946. He had a long time interest in aircraft and was one of the first to realize the extent to which the use of air power would alter warfare. The first combat model here was proposed in his 1916 book *Aircraft in Warfare* published during World War I. His book predicted that "the number of flying machines eventually to be utilized by ... the great military powers will be counted not by hundreds but by thousands ... and the issue of any great battle will be definitely determined by the efficiency of the aeronautical forces." This project investigates some of Lanchester's work on modeling combat.

Two forces numbering $x[t]$ and $y[t]$ are fighting each other. Lanchester assumed that if the forces were fighting a conventional war then the combat loss rates would be proportional to the total number of enemy forces. Expressed as a set of differential equations, this becomes

$$\frac{dx}{dt} = -ay$$
$$\frac{dy}{dt} = -bx$$

where a and b are positive constants representing the effectiveness of the y and x forces. Initial conditions would represent the initial strength of each force.

HINT 34.1. *The Basic Combat Model*
Begin your project with an explanation of the way these differential equations model combat. In particular, explain why the parameters a and b represent the effectiveness of the armies.

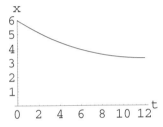 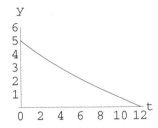

FIGURE 34.2: A Solution of the Combat Equations

Figure 34.2 illustrates the course of combat between two forces of equal effectiveness, $a = b = 0.1$, and different initial strengths. One force starts at 60,000 and the other at 50,000. After 12 units of time, the smaller force is annihilated and the larger force loses about 25,000 men.

You need to be careful with numerical results for these equations. Negative values of the variables won't make any sense to the model. With this in mind, perform some experiments to get a feel for the behavior of the model.

HINT 34.3. *Explicit Experiments*

Use the computer program **AccDEsol** to solve these differential equations for a larger value of **tfinal** > 12. ($a = b = 0.1$ and $x_0 = 60000$, $y_0 = 50000$.) Explain the odd behavior of the solutions for larger values of t. At what point do the numerical results fail to make sense in the combat model?

Experiment with different values of the initial troop strengths, x_0 and y_0. What happens if the two forces are of equal effectiveness and equal strength?

Experiment with various values of the parameters a and b.

You need to organize the results of your experiments with the explicit time solutions of the equations. Hint 34.3 should give you some ideas about the behavior of the model but at the same time not be easy to summarize. It will be helpful to take another view of the solutions that makes the comparison more natural by plotting both forces on one graph.

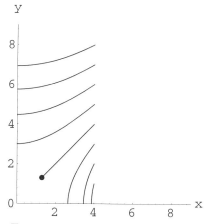

FIGURE 34.4: The Flow of Battle

HINT 34.5. *Flow Experiments*

The differential equations of the basic combat model are autonomous. Why? This means that you can use the **Flow2D** computer program to experiment with the behavior of your model. Modify that program to solve the equations with $a = b = 0.1$ and use all the initial conditions that you experimented with in the explicit solution experiments (Hint 34.3) in a single flow. At the very least, this will give you a way to compare all the results of those experiments at once.

Experiment with the flow solution of the basic combat model when $a \neq b$, using the same initial troop strengths as before.

Compare the results obtained by explicit solution with the flow animations.

A basic kind of comparison is to compute with a fixed-size x force and different initial sizes of the y force. You could make a list of initial conditions, for example, using *Mathematica's* Table[.] command:
    initconds = Table[{ 40000,y} , { y,0,80000, 10000}]

34.1. The Principle of Concentration

If one force can concentrate its entire strength on a portion of an opposing army and destroy it, it may then be able to attack the remaining portion of the opposing army with its remaining strength and destroy it. Lanchester believed air power would be so effective because of this. An air force can be deployed in full force against select portions of an opposing army.

HINT 34.6. *Divide and Conquer*
*1) First consider two forces x and y of equal effectiveness. Suppose the x-force starts with 50,000 men and the y-force with 70,000 men. Use the **rkSoln** or the **Flow2D** program to determine the outcome of the battle if all 50,000 x-troops faced all 70,000 y-troops.*
2) Let the x and y forces have equal effectiveness and let the x-force start at 50,000 and the y-force start at 70,000. Suppose now, however, that the x-force can concentrate its entire strength on 40,000 opposing troops before facing the remaining 30,000 y-force men in a second battle. How many men are left at the end of the first battle? What is the outcome of the second battle in this instance?
3) Suppose now that an x-force of 50,000 troops is outnumbered by 120,000 y-troops. Assuming that the x-force can strike a portion of the y-force in full strength and then face the other portion with its remaining men, is there any way for the x-force to win the battle? Justify your answer.

It would be convenient to have a simple analytical method to use to compute answers to questions like those in the exercise above. Computer simulations can probably give you the answers you need, but there is an easier way

34.2. The Square Law

Consider the case where the two forces have differing combat effectiveness. (In particular, we will use $a = 0.0106$ and $b = 0.0544$ for experiments. These coefficients have been used to model the course of the battle of Iwo Jima in World War II.) The outcome of the battle in this case is not as clear as with the battles above. A force of superior numbers may still lose to a smaller force if the combat effectiveness of the smaller force is high. We can still determine the outcome of a battle, however, by using the differential equations above to derive an integral invariant similar to the energy in a spring or the epidemic invariant or the predator-prey invariant from the CD text Section 24.4. Note that since

$$\frac{dx}{dt} = -ay$$
$$\frac{dy}{dt} = -bx$$

we can take the quotient of $\frac{dy}{dt}$ and $\frac{dx}{dt}$ to obtain

$$\frac{dy}{dx} = \frac{-bx}{-ay}$$

Now we can separate variables and integrate to get a relationship between x and y. (Review the invariants in CD Section 24.4 of the core text.)

HINT 34.7. *The Combat Invariant*
Perform the integration above, remembering that you will introduce an arbitrary constant while integrating. The solutions $x[t]$ and $y[t]$ will stay on the integral invariant for all time. Prove this once you have computed the formula as in CD Section 24.4.

The initial values of x and y will determine the value of the arbitrary constant introduced by the integration. What kind of curves are the integral invariants? For given values of a and b and initial strengths of the x and y forces, how can you use the integral invariant to determine who will win the battle?

Once you have found the invariant formula, you can derive Lanchester's "square law." What is the outcome of the battle if the arbitrary constant you introduced by integrating is 0? What is the outcome if the constant is negative? positive?

HINT 34.8. *The Square Law*
Complete the following sentences and derive the square law:

(1) *If $bx_0^2 < ay_0^2$, then ....*
(2) *If $bx_0^2 = ay_0^2$, then ....*
(3) *If $bx_0^2 > ay_0^2$, then ....*

HINT 34.9. *Consider the specific case of $a = 0.0106$ and $b = 0.0544$. Use the **Flow2D** program to solve the differential equations for several initial conditions. In particular, include an initial condition for which the two forces annihilate each other. Also include initial conditions for which the x force wins and the y force wins. What region of the x and y plane do the initial conditions have to be in for the x force to win? for the y force to win? for the forces to mutually annihilate each other?*
What is the connection between the square law and the flow?

34.3. Guerrilla Combat

Lanchester's combat equations have been modified to model combat in which one force is a conventional force and one force is a guerrilla force. The modification takes into account the fact that the fighting effectiveness of a guerrilla force is due to its ability to stay hidden. Thus, while losses to the conventional force are still proportional to the number of guerrillas, the losses to the guerrilla forces are proportional to both the number of the conventional force and its own numbers. Large numbers of guerrillas cannot stay hidden and, therefore, derive little advantage from guerrilla combat.

If x is a guerrilla force and y a conventional force, the modified combat equations are:

$$\frac{dx}{dt} = -axy$$
$$\frac{dy}{dt} = -bx$$

where here again a and b are constants representing the combat effectiveness of each force.

HINT 34.10. *Just as above, we can calculate an integral invariant to determine how solutions to the differential equation behave. Take the ratio of $\frac{dy}{dt}$ and $\frac{dx}{dt}$ to obtain $\frac{dy}{dx}$ and integrate to obtain an expression involving x and y and an arbitrary constant. What is the shape of these integral invariants? Derive another combat law:*

(1) *If ..., then the guerrilla force wins.*
(2) *If ..., then the forces annihilate each other.*
(3) *If ..., the conventional force wins.*

HINT 34.11. *Let $a = 0.0000025$ and $b = 0.1$. Use the **Flow2D** program to solve the differential equations for several initial conditions. Again include at least one initial condition for which the guerrilla force wins, at least one for which the conventional force wins, and one for which the forces annihilate each other. What region of the x and y plane do the initial conditions have to be in for the conventional force to win? for the guerrilla force to win? for the forces to annihilate each other?*

HINT 34.12. *Consider the case in which both forces are guerrilla forces. Then the combat equations become:*

$$\frac{dx}{dt} = -axy$$
$$\frac{dy}{dt} = -bxy$$

*Find the integral invariants in this case (they are the simplest of the three models you've seen). What do the invariants represent? Finally, derive another combat law that will enable you to predict the outcome of a battle given certain initial conditions. Use the **Flow2D** program to demonstrate your combat law.*

34.4. Operational Losses (Optional)

Our final modifications of Lanchester's combat model will take into account operational losses. These are losses that occur through accidents, "friendly" fire, or disease. We will assume that the rate of these losses is proportional to the size of the force. The new combat model then takes the form:

$$\frac{dx}{dt} = -ex - ay$$
$$\frac{dy}{dt} = -bx - fy$$

where a and b are the usual combat effectiveness constants and e and f are operational loss rates. It seems plausible that these should be small in comparison to a and b. This is a linear system like the ones we studied in the text.

HINT 34.13. *Suppose that a, b, e, and f are positive. Show that the characteristic roots of the linear dynamical system above are real and of opposite sign.*

In this instance, it is not possible to find an integral invariant as above. Notice, however, that the expressions describing rates of change are linear in x and y. If this is the case, it can be shown that if e and f are small compared to a and b then the solutions to these equations behave exactly like the solutions to the combat model without operational losses. You showed above that if the initial conditions are on a certain line through the origin, then the forces mutually annihilate each other. A unit vector that is drawn in the direction of this line is called a *characteristic vector* (*eigenvektor* in German). The key to extending the combat law to this case is in finding an invariant direction or unit vector in the direction.

The computer can find these vectors and enable you to derive a combat law for the model with operational losses. To do this we first write the differential equation in matrix form.

$$\begin{bmatrix} \frac{dx}{dt} \\ \frac{dy}{dt} \end{bmatrix} = \begin{bmatrix} -e & -a \\ -b & -f \end{bmatrix} \begin{bmatrix} x \\ y \end{bmatrix}$$

For example, in *Mathematica* we can define a matrix in the computer as a list of lists,

$$m = \{\{-e, -a\}, \{-b, -f\}\}$$

The command **Eigenvectors**[m] will then return two unit vectors. One will be in the wrong quadrant, but the other will give the direction of the line along which the initial conditions must lie in order for the two forces to mutually annihilate each other. For example, if $a = b = 0.1$ and $e = f = 0.001$, then the two eigenvectors are given in the list of lists,

$$\{\{0.707107, -0.707107\}, \{0.707107, 0.707107\}\}$$

The second list is the vector

$$\mathbf{E}_2 = \begin{bmatrix} 1/\sqrt{2} \\ 1/\sqrt{2} \end{bmatrix}$$

which points along the line "$y = x$" in the first quadrant.

The computer also has a command **Eigenvalues**[m] that returns the characteristic roots of the linear dynamical system. These are the same roots we studied in Chapter 23 of the core text,

$$\det \begin{vmatrix} -e - r & -a \\ -b & -f - r \end{vmatrix} = (e + r)(f + r) - ab = r^2 + (e + f)r - ab = 0$$

The roots are approximately $r_1 = 0.099$ and $r_2 = -0.101$ in this case. The single command **Eigensystem**[m] returns the eigenvalues and eigenvectors.

Eigenvectors and eigenvalues are related by the equation

$$\mathbf{mE}_2 = r_2 \mathbf{E}_2$$

or

$$\begin{bmatrix} -e & -a \\ -b & -f \end{bmatrix} \begin{bmatrix} x_0 \\ y_0 \end{bmatrix} = r_2 \begin{bmatrix} x_0 \\ y_0 \end{bmatrix}, \quad \text{for} \quad \mathbf{E}_2 = \begin{bmatrix} x_0 \\ y_0 \end{bmatrix}$$

In particular,

$$\begin{bmatrix} -1/1000 & -1/10 \\ -1/10 & -1/1000 \end{bmatrix} \begin{bmatrix} x_0 \\ x_0 \end{bmatrix} = -0.101 \begin{bmatrix} x_0 \\ x_0 \end{bmatrix}$$

for any constant x_0. In this case,

$$\begin{bmatrix} x[t] \\ y[t] \end{bmatrix} = ke^{-0.101t} \begin{bmatrix} 1/\sqrt{2} \\ 1/\sqrt{2} \end{bmatrix} = e^{-0.101t} \begin{bmatrix} x_0 \\ y_0 \end{bmatrix}$$

is a solution of the differential equations for any initial vector

$$\begin{bmatrix} x_0 \\ y_0 \end{bmatrix} = k \begin{bmatrix} 1/\sqrt{2} \\ 1/\sqrt{2} \end{bmatrix}$$

In other words, if $x_0 = y_0$, then $x[t] = y[t] = x_0 e^{-0.101 t}$.

HINT 34.14.

(1) Compute the matrix product
$$\begin{bmatrix} -1/1000 & -1/10 \\ -1/10 & -1/1000 \end{bmatrix} \begin{bmatrix} k \\ k \end{bmatrix}$$
for any constant k.

(2) Show that the matrix equation
$$\mathbf{m} \mathbf{E}_2 = r_2 \mathbf{E}_2$$
holds with $r_2 = -0.101$ and $\mathbf{E}_2 = (1/\sqrt{2}, 1/\sqrt{2})$.

(3) Compute the derivatives of the components of the vector below and verify the equation
$$\frac{d}{dt} \begin{bmatrix} x_0 e^{r_2 t} \\ y_0 e^{r_2 t} \end{bmatrix} = r_2 \begin{bmatrix} x_0 e^{r_2 t} \\ y_0 e^{r_2 t} \end{bmatrix}$$

(4) When $x_0 = y_0$, use the computations above to show that
$$\begin{bmatrix} -1/1000 & -1/10 \\ -1/10 & -1/1000 \end{bmatrix} \mathbf{X} = \frac{d\mathbf{X}}{dt} \quad \text{for the solution} \quad \mathbf{X} = \begin{bmatrix} x_0 e^{r_2 t} \\ x_0 e^{r_2 t} \end{bmatrix}$$

(5) What condition on r_2 is needed so that the equations
$$\begin{bmatrix} x[t] \\ y[t] \end{bmatrix} = \begin{bmatrix} x_0 e^{r_2 t} \\ y_0 e^{r_2 t} \end{bmatrix} = e^{r_2 t} \begin{bmatrix} x_0 \\ y_0 \end{bmatrix}$$
are parametric equations for a ray point toward the origin? In other words, when will the equations trace the line segment starting at (x_0, y_0) and ending at $(0, 0)$ for $0 \le t < \infty$?

In general, if we have an eigenvector-eigenvalue pair for a matrix $\mathbf{m}$,
$$\mathbf{m} \mathbf{E}_2 = r_2 \mathbf{E}_2$$
then the vector
$$\mathbf{X}[t] = k e^{r_2 t} \mathbf{E}_2$$
satisfies two conditions:
$$\mathbf{m} \mathbf{X} = \mathbf{m}(k e^{r_2 t} \mathbf{E}_2) = k e^{r_2 t} \mathbf{m} \mathbf{E}_2 = k e^{r_2 t} r_2 \mathbf{E}_2 = r_2 \mathbf{X}$$
and
$$\frac{d\mathbf{X}}{dt} = \frac{d(k e^{r_2 t})}{dt} \mathbf{E}_2 = r_2 k e^{r_2 t} \mathbf{E}_2 = r_2 \mathbf{X}$$
so that
$$\frac{d\mathbf{X}}{dt} = \mathbf{m} \mathbf{X}$$
If the eigenvalue r_2 is negative, then the function $\mathbf{X}[t]$ is a parametric form of the segment from (x_0, y_0) to $(0, 0)$. In other words, the solution tends to the origin along a line in the direction of $\mathbf{E}_2$.

HINT 34.15. *Consider the model above for two forces of equal combat effectiveness, $a = b = 0.1$, and operational loss rates $e = 0.00001$ and $f = 0.0001$. Use the Eigensystem command to find the eigenvector - eigenvalue pairs of the matrix above. Once you have the direction of the line of mutual annihilation, use the **Flow2D** program to plot initial conditions for which the x force wins, for which the y force wins, and for which the forces annihilate one another. What is the combat law for these particular parameters?*

How large can the operational losses become before the eigenvector associated with the negative eigenvalue no longer lies in the first quadrant? Is the model useful if this happens?

PROJECT 35

Drug Dynamics and Pharmacokinetics

Differential equations are useful in studying the dynamics of a drug in the body. The study of such dynamics is called "pharmacokinetics." Here is a basic example. Suppose a drug is introduced into the blood stream, say by an intravenous injection. The injection rapidly mixes with the whole blood supply and produces a high concentration of the drug everywhere in the blood. Several tissues will readily absorb the drug when its concentration is higher in the blood than in the tissue, so the drug moves into this second tissue "compartment." If this were the whole story, the concentration would eventually balance out so that the concentration in blood and tissue were both equal to the total amount of drug divided by the total volume. However, this usually is not the end of the story.

The kidneys remove the drug from the blood at a rate proportional to the blood concentration. This causes the blood concentration to drop, and eventually it drops below the tissue concentration. At that point, the drug flows from tissue back into the blood and is continually eliminated from the blood by the kidneys. In the long term, the drug concentration tends to zero in both blood and tissue. The speeds with which these various things happen is the subject of pharmacokinetics.

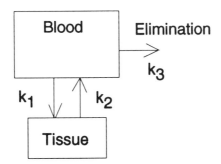

FIGURE 35.1: two-compartments

Why should we care about such dynamics? Some drugs have undesirable, or even dangerous, side effects if their concentration is too high. At the same time those drugs must be above a certain concentration to be effective for their intended use. As the drug is eliminated from the body, doses need to be given periodically in order to maintain the threshold level for effectiveness, yet doses cannot be too frequent or too large or the concentration will exceed a dangerous level.

This project studies the basic dynamics of the two-compartment drug model. Projects 15 and 16 study some important consequences of these dynamics. The "two-compartment model" corresponds to the previous story. More complicated dynamics might include an intramuscular injection that first diffuses into the blood stream, then into perfuse tissue, and finally is eliminated by the kidneys. Another dynamic might be a drug that is metabolized while it is in the tissue compartment.

The actual "compartments" can vary according to the drug. For example, "blood volume" often includes highly perfused lean tissue such as the heart, lungs, liver, kidney tissue, endocrine glands and occasionally the brain and spinal system (since these often present different barriers to drugs). In these cases, "tissue volume" consists of muscle and skin. Another possible "compartment" is the fat group, adipose and marrow. The anticoagulant warfarin (used as a blood thinner and in rat killer) becomes bound to a protein in the blood. It is not absorbed in tissue, but the bound state forms a second "compartment."

For now we just consider a drug that is introduced into the blood, diffuses into, tissue and is eliminated by the kidneys. Some of the drugs that fit the two-compartment model are aspirin (acetylsalicylic acid); creatinine, a metabolite of creatine produced by muscle contraction or degeneration; aldosterone; griseofulvin (an antifungus drug); and lecithin.

PRIMARY VARIABLES OF THE MODEL

The primary quantity we wish to measure is the drug concentration in the blood. That concentration is a function of time and is quickly and directly affected by the concentration of the drug in the tissue.

t – Time, beginning with the initial introduction of the drug, in hours. hr
$c_B = c_B[t]$ – The concentration of the drug in the blood in milligrams per liter. mg/l
$c_T = c_T[t]$ – The concentration of the drug in the tissue in milligrams per liter. mg/l

HINT 35.2. *Show that the units of the drug concentrations in mg/l are equal to units of micrograms per milliliter, so that we might measure the number of micrograms in a milliliter of a patient's blood, rather than the number of milligrams in a liter of the blood...*

Each drug and each patient have certain important constants associated with them. In the complete story these will have to be measured.

PARAMETERS OF THE MODEL

v_B - The volume of the blood compartment in liters. l
v_T - The volume of the tissue compartment in liters. l
k_1 - A rate constant for diffusion into the tissue compartment.
k_3 - A rate constant for elimination by the kidneys.

Concentration is an amount per unit volume. Suppose that a patient has 2.10 liters of blood and that 0.250 grams of a substance is introduced into his blood. When mixed, the concentration becomes $0.25 \times 1000/2.10 = 119$. (mg/l).

HINT 35.3. *Secondary Variables*
Give a general formula to convert the blood and tissue concentrations into the actual total amounts of the drug in the blood or tissue at a given time,

$a_B = a_T[t] = ??c_B[t]$ - The amount of the drug in the blood compartment. units?
$a_T = a_T[t] = ??c_T[t]$ - The amount of the drug in the tissue compartment. units?

Suppose a patient has $v_B = 2.10$ and $v_T = 1.30$. If $c_B[t]$ and $c_T[t]$ were known functions, what would the amounts $a_B[t]$ and $a_T[t]$ be?

35.1. Derivation of the Equations of Change

The primary dynamic mechanism for moving the drug into tissue is a "concentration gradient." If the blood concentration is higher than the tissue concentration, there is a flow of the drug from the blood into the tissue. The simplest relationship would be a linear one,

Rate of reduction of blood concentration into tissue $\propto (c_B - c_T)$

We use the parameter k_1 for the constant of proportionality. Suppose that this patient has no kidneys, so there is no elimination (or metabolism) of the drug. Then we have the rate of decrease of c_B equal to the negative derivative,

$$\frac{dc_B}{dt} = -k_1(c_B - c_T) = -k_1 c_B + k_2 c_T \qquad \text{(no kidneys)}$$

HINT 35.4. *What are the units of $\frac{dc_B}{dt}$? What does this force the units of k_1 to be?*

The drug that flows from blood to tissue is neither created nor destroyed in our model. We need to account for this **amount** in the equation for the rate of change of the tissue concentration, $\frac{dc_T}{dt}$. Consider the case of our hypothetical patient with a blood volume of 2.10 liters and tissue volume of 1.30 liters. If c_B is reduced by 10.0 mg/l in an hour, then the amount of drug that has left in the hour is $10.0 \times 2.10 = 21.0$ (Note units: $mg/l \times l = mg$). This raises the amount of drug in the tissue by 21 mg, but the concentration in the tissue goes up by $21/1.30 = 16.$ mg/l. Notice the units in our calculations and use them in the next exercise.

HINT 35.5. *What is the rate of change of the amount of drug in the blood, $\frac{da_B}{dt}$, in terms of patient parameters and $\frac{dc_B}{dt}$?*

What is the rate of change of the amount of drug in the tissue, $\frac{da_T}{dt}$, in terms of patient parameters and $\frac{dc_T}{dt}$?

Show that
$$\frac{dc_T}{dt} = k_2(c_B - c_T) = k_2 c_B - k_2 c_T$$

where $k_2 = k_1 v_B/v_T$.

Finally, we shall hypothesize a linear elimination law for a patient with kidneys, that is,

Rate of reduction of blood concentration by kidneys $\propto c_B$

so that the two terms reducing blood concentration are

$$\frac{dc_B}{dt} = -k_1(c_B - c_T) - k_3 c_B$$

HINT 35.6. *The Dynamics*
Begin your project with an explanation of the pharmacological meaning of the equations:

$$\frac{dc_B}{dt} = -(k_1 + k_3)\, c_B + k_1\, c_T$$
$$\frac{dc_T}{dt} = k_2\, c_B - k_2\, c_T$$

Give the units of your variables and parameters and explain why the parameters v_B, v_T, k_1 and k_3, but not k_2, need to be measured.

Now we want to begin with some numerical experiments to get a feel for the model.

HINT 35.7. *Numerical Experiments with No Kidneys*
Modify the **AccDEsol** program to find some explicit solutions of the concentration equations. Your initial experiment might look as follows:

```
f := -(k1 + k3) cB + k1 cT;
g := k2 cB - k2 cT;
k1 = 0.17;
k2 = k1 vB/vT;
k3 = 0.0 (*No kidney function.*);
ti = 0.0 (*Initial time.*);
cBi = 47.7 (*Initial blood concentration.*);
cTi = 0.0 (*Initial tissue concentration.*);
tfinal = 10 ;
dt = .1 ;
AccDEsol[{1,f,g},{t,cB,cT},{ti,cBi,cTi},tfinal,dt]; (*Use the AccDEsol pgm.*)
```

Once you have your program working properly, compute what happens to a patient with these parameters after 48 hours. Your graphs should look like Figure 35.8.

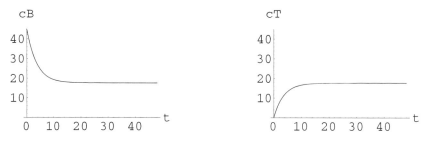

FIGURE 35.8: Forty-eight Hours without Kidneys

HINT 35.9. *Equilibrium without Kidneys*
A patient with no kidneys, $v_B = 2.10$, $v_T = 1.30$, and $k_1 = 0.17$ is administered 100 mg of a drug that is not metabolized. What will his blood concentration be after 48 hours? (Hints: What is the total amount of drug in his body? What is the total volume? What is the connection between amount and concentration?)

A patient is administered an initial amount a_I of drug. Show that the initial blood concentration is $c_I = a_I/v_B$. Let c_∞ be an unknown amount for the equilibrium concentration.

Show that the equilibrium amount of drug in the blood is $c_\infty v_B$ and the equilibrium amount of drug in tissue is $c_\infty v_T$, for a total $(v_B + v_T) c_\infty$. Set this equal to a_I and solve for $c_\infty = ?$

The graphs above show that c_B and c_T both tend to a limit as t tends to infinity,

$$\lim_{t \to \infty} c_B[t] = \lim_{t \to \infty} c_T[t] = c_\infty$$

Why do both variables tend to this limit? What is a formula for this limiting concentration, c_∞, in terms of patient parameters and initial dose?

Your next task is to view these solutions from the point of view of the flow solutions. This lets us see many initial conditions at once.

HINT 35.10. *The Flow Picture without Kidneys*
*Modify the **Flow2D** program to check a number of equilibria, but with many initial doses. Your program might look as follows:*

```
f := -(k1 + k3) cB + k1 cT;
g := k2 cB - k2 cT;
k1 = 0.17;
k2 = k1 vB/vT;
k3 = 0.0 (*No kidney function.*);
initDoses = Table[{c,0},{c,0,100,5}]
initConcs = initDoses/vB;
tfinal = 48 ;
dt = .1 ;
numgraphs = 10 ;
flow2D[ {f,g}, {cB,cT}, initConcs, tfinal, dt, numgraphs ];
```

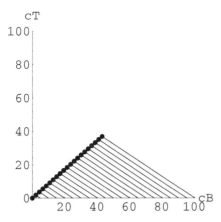

FIGURE 35.11: No Kidney Equilibria for Various Doses

HINT 35.12. *Conjecture*
Describe the flow animation of the patient with no kidneys. What are the apparent paths of the solutions? What is the locus of points where the solutions end up? Is there more than one equilibrium point?

The solutions of the flow animation are surprisingly simple and yet not "typical" of the main examples studied in Chapters 23 and 24 of the core text. However, what you see is indeed correct. Things do flow along parallel lines toward another line of equilibria in the phase plane.

HINT 35.13. *Symbolic Proofs for a Degenerate System*

(1) *Show that the constant solutions $c_B[t] = c_\infty = c_T[t]$, for any constant c_∞, are equilibrium points of the dynamical system corresponding to no kidneys:*

$$\frac{dc_B}{dt} = -k_1 c_B + k_1 c_T$$
$$\frac{dc_T}{dt} = k_2 c_B - k_2 c_T$$

(2) *Show that the following is an analytical solution to our model:*

$$\begin{bmatrix} c_B[t] \\ c_T[t] \end{bmatrix} = \begin{bmatrix} c_\infty \\ c_\infty \end{bmatrix} + c_1 \, e^{-ht} \begin{bmatrix} k_1 \\ -k_2 \end{bmatrix}$$

where c_1 and c_∞ are constant and $h = k_1 + k_2$. In other words, plug the solutions $c_B[t] = c_\infty + c_1 k_1 \, e^{-ht}$ and $c_T[t] = c_\infty - c_1 k_2 \, e^{-ht}$ into the equations and verify that they do solve them. (Hint: $c_B - c_T = (k_1 + k_2) c_1 \, e^{-(k_1+k_2)t}$.)

(3) *Show that these analytical solutions satisfy $\lim_{t \to \infty} c_B[t] = \lim_{t \to \infty} c_T[t] = c_\infty$.*

(4) *Review the derivation of the equation for a vector parametric line $\mathbf{X} = \mathbf{P} + t\mathbf{D}$ (Chapter 17 of the core text) and use that idea to show that the solution*

$$\mathbf{C}(t) = \mathbf{C}_\infty + e^{-ht} \, \mathbf{K}$$

$$\begin{bmatrix} c_B[t] \\ c_T[t] \end{bmatrix} = \begin{bmatrix} c_\infty \\ c_\infty \end{bmatrix} + e^{-ht} \begin{bmatrix} c_1 k_1 \\ -c_1 k_2 \end{bmatrix}$$

gives the motion along the parallel lines to the other line that you conjectured in the previous exercise. Notice that the scalar coefficient e^{-ht} starts at 1 and decreases to 0 as $t \to \infty$.

(5) *If the initial dose is the amount a_I, so $c_B[0] = a_I/v_B$ and $c_T[0] = 0$, show that $c_\infty = a_I/(v_B + v_T)$ and the constant c_1 from part (2) must be $c_1 = \frac{a_I}{k_1} \frac{v_T}{v_B} \frac{1}{v_B + v_T}$.*

In summary, show that the unique solution to the drug equations without kidneys with initial condition $(c_I, 0)$ may be written in terms of the initial concentration, c_I, and parameters as

$$c_B[t] = c_I \left(u_1 \, e^{-ht} + u_2 \right)$$
$$c_T[t] = c_I \left(-u_2 \, e^{-ht} + u_2 \right)$$

where

$$u_2 = \frac{v_B}{v_B + v_T} \quad \text{and} \quad u_1 = \frac{v_T}{v_B + v_T}$$

At this point we have a complete numerical and analytical solution to the dynamics of a patient without kidneys. This is a mathematical start but not the main idea we want to explore. However, it is a useful start, because elimination is typically a slower process than tissue absorption. You will see some similarity in the solutions with a small k_3 compared with $k_3 = 0$.

HINT 35.14. *Numerical Experiments with Kidneys*
Modify the **AccDEsol** and **Flow2D** programs to solve the drug equations when $k_3 \neq 0$. For example,

f := -(k1 + k3) cB + k1 cT;
g := k2 cB - k2 cT;

k1 = 0.17;
k2 = k1 vB/vT;
k3 = 0.03;

ti = 0.0 (*Initial time.*);
cBi = 45.0 (*Initial blood concentration.*);
cTi = 0.0 (*Initial tissue concentration.*);

tfinal = 72 ;
dt = 0.5 ;

AccDEsol[{1,f,g},{t,cB,cT},{ti,cBi,cTi},tfinal,dt]; (*Using the AccDEsol pgm.*)

initConcs = Table[{c,0},{c,0,100,5}];

numgraphs = 10 ;
flow2D[{f,g}, {cB,cT}, initConcs, tfinal, dt, numgraphs];

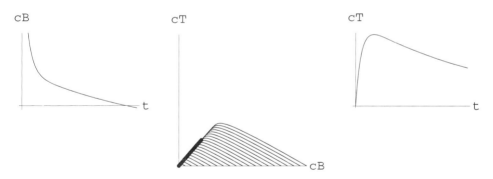

FIGURE 35.15: Seventy-two Hours with Kidneys

HINT 35.16. *Equilibrium with Kidneys*
If $k_1 > 0$ and $k_3 > 0$, prove that $(0,0)$ is the only equilibrium for the drug equations. Explain intuitively why this equilibrium must be an attractor.

Notice that our concentration equations have the form of a linear system

$$\frac{dc_B}{dt} = -(k_1+k_3)\,c_B + k_1\,c_T \qquad \frac{dx}{dt} = a_x\,x + a_y\,y$$
$$\frac{dc_T}{dt} = k_2\,c_B - k_2\,c_T \qquad \frac{dy}{dt} = b_x\,x + b_y\,y$$

where $a_x = -(k_1 + k_3)$, $a_y = k_1$, $b_x = k_2$, and $b_y = -k_2$. This implies that the concentration solutions may be written in terms of linear combinations of the exponential functions $e^{-h_1 t}$ and $e^{-h_2 t}$ where $-h_1$ and $-h_2$ satisfy the characteristic equation:

$$\det \begin{vmatrix} a_x - r & a_y \\ b_x & b_y - r \end{vmatrix} = (a_x - r)(b_y - r) - a_y b_x = r^2 - (a_x + b_y) r + (a_x b_y - a_y b_x) = 0$$

HINT 35.17. *The Analytical Form of the Solutions*

(1) Use Theorem 24.1 from Chapter 24 of the core text to show that the solutions $c_B[t]$ and $c_T[t]$ may be written as a linear combination of $e^{-h_1 t}$ and $e^{-h_2 t}$,

$$c_B[t] = b_1 e^{-h_1 t} + b_2 e^{-h_2 t}$$

where

$$h_1 = \frac{1}{2}[(k_1 + k_2 + k_3) + \sqrt{(k_1 + k_2 + k_3)^2 - 4 k_2 k_3}]$$

and

$$h_2 = \frac{1}{2}[(k_1 + k_2 + k_3) - \sqrt{(k_1 + k_2 + k_3)^2 - 4 k_2 k_3}]$$

(2) Use the Local Stability Theorem from Chapter 24 to prove that $(0,0)$ is an attractor when $k_j > 0$, for $j = 1, 2, 3$. (Typically, $k_3 \ll k_1$, although this is not needed for your proof.)
(3) Show that

$$h_1 h_2 = k_2 k_3 \quad \text{and} \quad h_1 + h_2 = k_1 + k_2 + k_3$$

(4) What happens to h_1 and h_2 in the case of nonfunctioning kidneys ($k_3 = 0$)?
(5) Let $k_1 = 0.17$, $v_B = 2.10$, $v_T = 1.30$. Compute h_1 and h_2 in two cases: $k_3 = 0.03$ and $k_3 = 0.0$.

What we know from the theory is that

$$c_B[t] = b_1 e^{-h_1 t} + b_2 e^{-h_2 t}$$

for some constants b_1 and b_2, where h_1 and h_2 are computed from the parameters by the quadratic formula of the previous exercise. The exact values of b_1 and b_2 are given in Hint 35.20. The details are messy but amount to some formulas that the computer can easily compute for us.

Reasonable approximations for b_1 and b_2 are easy to guess in the case $h_1 \gg h_2$. If the fast absorption into the tissue is so fast that the kidneys do nothing during this period, we know that the solution looks like $c_B[t] = b_1 e^{-h_1 t} + c_\infty$ where

$$b_1 = c_I \frac{v_T}{v_B + v_T} \quad \text{and} \quad c_\infty = c_I \frac{v_B}{v_B + v_T}$$

HINT 35.18. *Justify this last claim, $c_B[t] = b_1 e^{-h_1 t} + c_\infty$, for the constants above. What are the necessary assumptions? (Hint: See a previous exercise.)*

Show that $k_1 \gg k_3$ implies that $h_1 \gg h_2$ for some reasonable meaning of $\gg$. (If you want to be formal, take k_1 (and k_2) positive but not infinitesimal and $0 < k_3 \approx 0$. Show that $h_1 \approx k_1 + k_2$, while $h_2 \approx 0$. Connect your formalization with the real problem.)

In the case where $k_3 = 0$, the full solution could be written

$$\begin{bmatrix} c_B \\ c_T \end{bmatrix} = c_1 \ e^{-ht} \begin{bmatrix} k_1 \\ -k_2 \end{bmatrix} + \frac{c_\infty}{k_1} e^{0t} \begin{bmatrix} k_1 \\ k_1 \end{bmatrix}$$

The actual exact symbolic solution when $k_3 \neq 0$ can be written in the vector form

$$\begin{bmatrix} c_B \\ c_T \end{bmatrix} = c_1 \ e^{-h_1 t} \begin{bmatrix} k_1 \\ h_2 - k_2 \end{bmatrix} + c_2 \ e^{-h_2 t} \begin{bmatrix} k_1 \\ h_1 - k_2 \end{bmatrix}$$

HINT 35.19. *Show that if $k_3 \approx 0$, the two solutions immediately before this exercise are close to one another for finite time.*

If we let

$$c_1 = \frac{c_I}{k_1} \frac{h_1 - k_2}{h_1 - h_2} \quad \text{and} \quad c_2 = \frac{c_I}{k_1} \frac{k_2 - h_2}{h_1 - h_2}$$

then we get the initial conditions

$$\begin{bmatrix} c_I \\ 0 \end{bmatrix} = \frac{c_I}{k_1} \frac{h_1 - k_2}{h_1 - h_2} \begin{bmatrix} k_1 \\ h_2 - k_2 \end{bmatrix} + \frac{c_I}{k_1} \frac{k_2 - h_2}{h_1 - h_2} \begin{bmatrix} k_1 \\ h_1 - k_2 \end{bmatrix}$$

HINT 35.20. *Show that this choice of constants c_1 and c_2 immediately before this exercise may be written in the form*

$$c_B[t] = c_I \left(w_1 \ e^{-h_1 t} + w_2 \ e^{-h_2 t} \right)$$
$$c_T[t] = c_I \left(-w_3 \ e^{-h_1 t} + w_4 \ e^{-h_2 t} \right)$$

where

$$w_1 = \frac{h_1 - k_2}{h_1 - h_2} \quad , \quad w_2 = \frac{k_2 - h_2}{h_1 - h_2}$$

$$w_3 = \frac{h_1 - k_2}{h_1 - h_2} \frac{k_2 - h_2}{k_1} = w_4 = \frac{k_2 - h_2}{h_1 - h_2} \frac{h_1 - k_2}{k_1}$$

These formulas give the exact symbolic unique solution of the full drug dynamics equation with initial condition $(c_I, 0)$

HINT 35.21. *Compare the approximate solution function $c_A[t] = u_1 \ e^{-h_1 t} + u_2 \ e^{-h_2 t}$ to the **AccDEsol** solution of the drug equations with initial concentration $c_I = 1$ and to the exact symbolic solution $c_B[t] = w_1 \ e^{-h_1 t} + w_2 \ e^{-h_2 t}$. (Use the values of the parameters k_1, k_3, etc., from the exercises above.) Calculate the associated constants u_1 versus w_1 and u_2 versus w_2, using the formulas from Exercise 35.13 and the previous exercise.*

What happens when $k_3 = 0$? What happens when k_3 is nearly as large as k_1?

```
k1 = 0.17;
vB = 2.1;
vT = 1.3;
k3 = 0.03;
k2 = k1 vB/vT;
h1 = ((k1 + k2 + k3) + Sqrt[(k1 + k2 + k3)^2 - 4 k2 k3])/2;
h2 = ((k1 + k2 + k3) - Sqrt[(k1 + k2 + k3)^2 - 4 k2 k3])/2;
u1 = vT/(vT + vB)
w1 = (h1 - k2)/(h1 - h2)
```

```
u2 = vB/(vT + vB)
w2 = (k2 - h2)/(h1 - h2)
cB[t_] := cI (w1 Exp[ -h1 t] + w2 Exp[ -h2 t]);
cB[t]
```

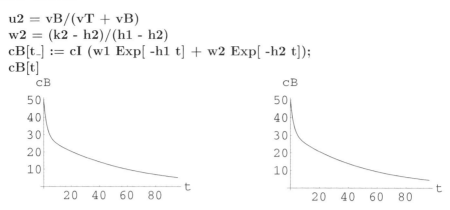

FIGURE 35.22: Computer Solution and Approximate Coefficients

35.2. Where Do We Go from Here?

The exact symbolic solution to a single dose [or initial conditions $(c_I, 0)$] given in the previous exercise is the starting point for the Project 15. That project explores things like the peak tissue concentration, the time interval during which the concentration remains effective, and the total integral of effective concentration. You could go to that project now to round out your thinking on pharmacokinetics.

Project 16 uses this single-dose symbolic solution and some basic work on logs from Project 15 to show how drug concentration data can be used to measure the rate parameters of this model. You might go to that project now instead of working Project 15.

Other alternatives for completing this project are to study more complicated dosing regimens. Usually a single dose of a drug is not given, but rather doses are either given periodically or a constant flow of drug is maintained in an intravenous fluid. The analysis of these two regimens follows as your third and fourth alternatives. Finally, a fifth alternative completion of the project is to study dosing by intramuscular injection. In this case, the drug must diffuse from the muscle into the blood before the two-compartment dynamics takes effect. This amounts to a three-compartment model.

35.3. Periodic Intravenous Injections (Optional Project Conclusion)

Suppose a patient is given a first dose of a drug, a_1 (mg) at time $t = 0$. Typically, this dose is larger than subsequent doses. After a time interval Δt, say 6 hours, a second dose a_2 is administered. We know that the exact solution from time $t = 0$ to $t = \Delta t$ is given by the exponential sum,

$$c_B[t] = c_I \left(w_1 e^{-h_1 t} + w_2 e^{-h_2 t} \right)$$
$$c_T[t] = c_I \left(-w_3 e^{-h_1 t} + w_4 e^{-h_2 t} \right)$$

where $c_I = a_1/v_B$ and

$$w_1 = \frac{h_1 - k_2}{h_1 - h_2} \quad , \quad w_2 = \frac{k_2 - h_2}{h_1 - h_2}$$

$$w_3 = \frac{h_1 - k_2}{h_1 - h_2} \frac{k_2 - h_2}{k_1} = w_4 = \frac{k_2 - h_2}{h_1 - h_2} \frac{h_1 - k_2}{k_1}$$

or we could simply use **AccDEsol** to find the solution.

At time $t = \Delta t$, the blood concentration is increased by a_2/v_B, but the old drug is not absent from the blood (and we often want to be sure it does not drop below a threshold of effectiveness). Say the concentration just before the second dose is $c_B[\Delta t]$ (using the formula above for $c_B[t]$.) Our new initial concentration becomes $e_1 = c_B[\Delta t] + a_2/v_B$.

Also, the concentration in the tissue is not zero at time Δt. We can use the formula above to compute $e_2 = c_T[\Delta t]$. As a result, we want to solve the new initial value problem

$$c_B[t_0 = \Delta t] = e_1$$
$$c_T[t_0 = \Delta t] = e_2$$
$$\frac{dc_B}{dt} = -(k_1 + k_3)\, c_B + k_1\, c_T$$
$$\frac{dc_T}{dt} = k_2\, c_B - k_2\, c_T$$

As an example, let us take $a_1 = 200$ *(mg)* and $a_2 = 100$ *(mg)* using the parameters $k_1 = 0.17$, $v_B = 2.10$, $v_T = 1.30$, $k_3 = 0.03$, as before. The values of concentration before the second dose are $c_B[6] = 52.6$ and $c_T[6] = 49.7$.

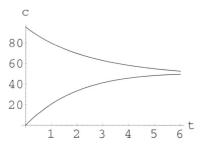

FIGURE 35.23: Concentrations up to Time 6

This makes the new initial value problem beginning with the second dose

$$c_B[t_0 = \Delta t] = 52.6. + 100/2.10 = 100.$$
$$c_T[t_0 = \Delta t] = 49.7$$
$$\frac{dc_B}{dt} = -(0.2)\, c_B + 0.17\, c_T$$
$$\frac{dc_T}{dt} = 0.275\, c_B - 0.275\, c_T$$

Notice that the initial value of c_T is no longer zero.

We can put this into **AccDEsol** as a new initial value problem and solve for six more time units, then repeat the process. Or we can find the exact symbolic solution to this initial value problem and compute the concentrations six units later, add the third dose and continue with that procedure. In either case we get the long-term drug dosages and the question is:

Will the concentrations build up? What is the long-term max? The min?

That's your project.

Here's some help if you want to use exact solutions which aren't really any better than numerical ones from AccDEsol. The general solution to the differential equations may be written in the vector form

$$\begin{bmatrix} c_B \\ c_T \end{bmatrix} = c_1 \, e^{-h_1 t} \begin{bmatrix} k_1 \\ h_2 - k_2 \end{bmatrix} + c_2 \, e^{-h_2 t} \begin{bmatrix} k_1 \\ h_1 - k_2 \end{bmatrix}$$

If we reset our clock at the second dose, we want to have

$$\begin{bmatrix} c_B[0] \\ c_T[0] \end{bmatrix} = \begin{bmatrix} e_1 \\ e_2 \end{bmatrix}$$

Notice that substitution of $t = 0$ into the general vector form yields

$$\begin{bmatrix} c_B[0] \\ c_T[0] \end{bmatrix} = c_1 \begin{bmatrix} k_1 \\ h_2 - k_2 \end{bmatrix} + c_2 \begin{bmatrix} k_1 \\ h_1 - k_2 \end{bmatrix} = \begin{bmatrix} e_1 \\ e_2 \end{bmatrix}$$

or in matrix form

$$\begin{bmatrix} k_1 & k_1 \\ h_2 - k_2 & h_1 - k_2 \end{bmatrix} \begin{bmatrix} c_1 \\ c_2 \end{bmatrix} = \begin{bmatrix} e_1 \\ e_2 \end{bmatrix}$$

For example, these equations can be solved with *Mathematica* by typing

    e1 = 100.;
    e2 = 49.7;
    {c1,c2} = LinearSolve[{{k1,k1},{h2 - k2 , h1 - k2}},{e1,e2}]
    cB[t_] := c1 k1 Exp[-h1 t] + c2 k1 Exp[-h2 t];
    cT[t_] := c1 (h2 - k2) Exp[-h1 t] + c2 (h1 - k2) Exp[-h2 t];
    cB[6]
    cT[6]

Our calculations with the parameters and doses above gave the following:

$c_B[6] = 52.6$ $c_B[12] = 71.2$ $c_B[18] = 89.7$ $c_B[24] = 107.$

$c_T[6] = 49.7$ $c_T[12] = 72.5$ $c_T[18] = 92.4$ $c_T[24] = 159.$

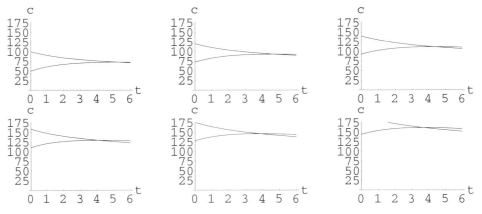

FIGURE 35.24: Concentrations between Doses 2 — 3, 3 — 4, 4 — 5, 5 — 6, 6 — 7, 7 — 8

$c_B[24] = 107.$ $c_B[30] = 123.$ $c_B[36] = 138.$ $c_B[42] = 152.$
$c_T[24] = 159.$ $c_T[30] = 128.$ $c_T[36] = 144.$ $c_T[42] = 159.$

What are the concentrations at the end of 48 hours, just before the ninth dose?

Do the concentrations continue to build up over time, or do they reach some maximum level and stop growing?

35.4. Steady Intravenous Flow

The previous problem started with an initial dose of 200 mg and then followed that with doses of 100 mg every 6 hours. **What would happen to the patient's drug concentrations if we constantly fed the drug into the blood stream at the rate of $100/6 = 16.7$ mg/hr?** This is much easier for us to analyze, because the "forcing term" or dosage is continuous.

First, we need to modify the basic equations,

$$\frac{dc_B}{dt} = -(k_1 + k_3) c_B + k_1 c_T$$
$$\frac{dc_T}{dt} = k_2 c_B - k_2 c_T$$

Now we have an additional term of a constant growth r added to the blood concentration. In the case described in the previous paragraph, r = 16.7/vB = 7.94 $mg/l/hr$. In general, this constant dose rate makes our initial value problem:

$$c_B[0] = 200/2.1 = 95.2$$
$$c_T[0] = 0$$
$$\frac{dc_B}{dt} = -(k_1 + k_3) c_B + k_1 c_T + r$$
$$\frac{dc_T}{dt} = k_2 c_B - k_2 c_T$$

HINT 35.25. *Solve this initial value problem (using the computer with **AccDEsol** or **Flow2D** or an exact method by changing variables) and show that its equilibrium level is*

$$(c_B, c_T) = \left(\frac{r}{k_3}, \frac{r}{k_3}\right)$$

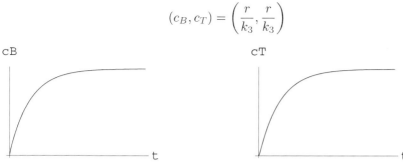

FIGURE 35.26: A Year of 16.7 $mg/l/hr$ Dosing

It seems reasonable to guess that the average concentration of the drug levels in the patient dosed every 6 hours (in the previous example) would build up to the equilibrium level of the continuously dosed patient. How high is this level?

35.5. Intramuscular Injection – a Third Compartment

Our final example of dosing is an intramuscular injection. In this case the drug is injected in high concentration in a muscle and then the drug diffuses into the blood. Once in the blood, it diffuses into the tissue compartment and is eliminated by the kidneys.

PARAMETERS FOR IM INJECTION

v_M - the volume of the injection site
k_4 - the rate parameter for diffusion into the blood
$k_5 = k_4 v_M / v_B$

HINT 35.27. *Explain the meaning of the system of differential equations:*

$$c_M[0] = 1000$$
$$c_B[0] = 0$$
$$c_T[0] = 0$$
$$\frac{dc_M}{dt} = -k_4 (c_M - c_B)$$
$$\frac{dc_B}{dt} = -(k_1 + k_3 + k_5) c_B + k_1 c_T + k5 c_M$$
$$\frac{dc_T}{dt} = k_2 c_B - k_2 c_T$$

in terms of the intramuscular injection. What is the amount of drug in the injection?

Use the computer to solve this system with $v_M = 0.250$ and $k_4 = 1.5$ and the parameters from the previous examples.

Chapter 12

Forced Linear Equations

One of the important traditional applications of differential equations is in understanding physical oscillators. Mechanical oscillators like the suspension of your car or electrical oscillators like the tuner in your radio often can be described by linear differential equations with an applied "force" such as the bumps in the road or a transmitted radio signal. These forces introduce time-dependent terms in the equations and make them nonautonomous. Still, the autonomous (or "homogeneous") solutions play an important role in understanding the forced oscillations. These "homogeneous" solutions are "transients" physically.

The first project in this chapter connects the physics of forcing, transients, and steady-state behavior with the mathematical solution of forced linear equations. The second project explains the buzz in your old car around 45 mph, and the third project shows how to filter out an unwanted radio signal.

PROJECT 36

Forced Vibration – Nonautonomous Equations

This is a project on linear initial value problems studied in engineering. We have studied some of that material in Chapter 23 of the main text. This project covers the most important basic aspects of forced linear oscillators: connecting the physics and math. This is interesting material even if you are not in engineering, but not as interesting as the applications such as the project on resonance or the notch filter described below. This project prepares you to separate the math of the "steady-state" and "transient" parts of the solution to those forced oscillator equations.

You should think of this project as a report to your engineering manager. She has been out of engineering school for a long time, so she may be rusty on the mathematical ideas. Be sure your report explains the quantities physically and mathematically.

HINT 36.1. *Preliminary*
Begin your projects with the derivation of the equation governing the mechanical or electrical oscillators associated with the second-order equations

$$a\frac{d^2x}{dt^2} + b\frac{dx}{dt} + cx = 0$$

(Choose one or the other for your whole project.) In the case of a mass m attached to a spring with Hooke's law constant s and suspended in a damping material of constant c, this is just the Newton's law "F = m A" derivation of the core text. If y measures the weight's displacement from equilibrium, my'' is the "m A" term, while the spring force is $-sy$ and the damping force is $-cy'$. These forces are negative to reflect the fact that if you extend the spring to $+y$, it pulls back, etc. Draw a complete figure for your report and do the algebra necessary to put things in the final form

$$my'' + cy' + sy = 0$$

Derivation of the equation for an electrical oscillator consists of using current-voltage laws for resistors, inductors, and capacitors, then combining them with Kirchhoff's voltage law. One case is described in the project on the notch filter.

36.1. Solution of the Autonomous Linear Equation

Your old e-homework program (core text Exercise 23.7.2) that solves the autonomous (also called homogeneous) initial value problems

$$my'' + cy' + sy = 0$$
$$y(0) = y_0$$
$$y'(0) = z_0$$

can be part of your project, but we want you to extend it to solve forced systems.

You should include a few paragraphs of explanation on how the program works and why the various cases arise by assuming an exponential solution. In other words, review the basic theory for your engineering manager, who may be a little rusty on these facts. Give examples where the system oscillates and where it does not. The computer program **SpringFriction** shows the result of lowering friction. You may use it or excerpts in your report as long as you explain what the excerpts mean.

HINT 36.2. *Give a careful explanation of the symbolic solution of the homogeneous problem above. Be sure to include the mathematics associated with the interesting physical phenomenon of the onset of oscillation as damping friction decreases, causing the mathematical phenomenon of complex roots.*

36.2. Transients – Limiting Behavior

The solutions to the autonomous linear IVP are called transients, because they tend to zero as $t \to \infty$. Transients don't stick around. In Exercise 23.6.2 of the text, you showed that the fact that the constant c in the differential equation is positive means that any solution to the homogeneous IVP tends to zero. (We also showed this by an energy argument in Section 24.4, though you will probably find it easier to explain in terms of explicit solutions.) Include the solution to this problem in your projects:

HINT 36.3. *If $y_T[t]$ satisfies:*

$$my'' + cy' + sy = 0$$

with m, c, and s positive, then $\lim_{t \to \infty} y_T[t] = 0$, no matter what initial conditions $y_T[t]$ satisfies.

This fact will play an important role in explaining the physical form of solutions to forced systems where the solutions "tend toward steady-state" as "the transients die out."

36.3. Superposition for the Spring System

Imagine applying an external force to the end of the spring, $F[t]$, besides the spring's restoring force and the friction $-cy'$. The resulting motion will be a solution to the nonautonomous differential equation

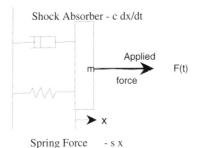

FIGURE 36.4: System with Applied Force

$$m y'' + c y' + s y = F[t]$$

The physical superposition principle says that "the response to a sum of forces is the sum of the responses to the separate forces." This verbal formulation is mostly right, but not quite precise enough to formulate mathematically. Mathematical superposition is needed to decompose our solutions into steady-state solutions and transients.

HINT 36.5. *Superposition of Linear Oscillators*
Suppose that $y_1[t]$ satisfies

$$m y'' + c y' + s y = F_1[t]$$

and $y_2[t]$ satisfies

$$m y'' + c y' + s y = F_2[t]$$

Show that $y[t] = y_1[t] + y_2[t]$ satisfies

$$m y'' + c y' + s y = F_1[t] + F_2[t]$$

Also prove that if $y_S[t]$ satisfies

$$m y'' + c y' + s y = F[t]$$

and $y_T[t]$ satisfies

$$m y'' + c y' + s y = 0$$

then $y[t] = y_S[t] + y_T[t]$ still satisfies

$$m y'' + c y' + s y = F[t]$$

The (one and only) motion of the spring system is actually the unique solution to an initial value problem. How can we formulate the superposition principle in terms of IVPs so that the mathematical solution is the physical motion? We either have to use special initial conditions or state the law in terms of steady-state solutions.

HINT 36.6. *Suppose that $y_1[t]$ satisfies*

$$y(0) = p_1$$
$$y'(0) = v_1$$
$$m y'' + c y' + s y = F_1[t]$$

and $y_2[t]$ satisfies

$$y(0) = p_2$$
$$y'(0) = v_2$$
$$m\,y'' + c\,y' + s\,y = F_2[t]$$

What initial conditions does the function $y[t] = y_1[t] + y_2[t]$ satisfy; that is, what are the values of p_3 and v_3 in the following if y is the solution?

$$y(0) = p_3$$
$$y'(0) = v_3$$
$$m\,y'' + c\,y' + s\,y = F_1[t] + F_2[t]$$

Include a mathematical formulation of the superposition principle in your report, explaining in very few words what this means mathematically. There are several ways to do this.

HINT 36.7. *Nonlinear Equations*
Show that the differential equation

$$\frac{dx}{dt} - \sqrt{x} = F[t]$$

does not satisfy the superposition principle. Notice that $x_1[t] = t^2$ satisfies $\frac{dx}{dt} - \sqrt{x} = t$, that $x_2[t] = t^2/4$ satisfies $\frac{dx}{dt} - \sqrt{x} = 0$, but that $x_1[t] + x_2[t]$ does not satisfy $\frac{dx}{dt} - \sqrt{x} = t + 0$.

36.4. Equations Forced by Gravity

We will begin with a very simple applied force, a constant. Suppose we hang our mass vertically on its spring and shock absorber (like a real car's suspension). The force due to gravity is $m\,g$ and Newton's "F = mA" law becomes

$$m\frac{d^2y}{dt^2} = -s\,y - c\frac{dy}{dt} - m\,g$$

when we measure y in an upward direction with $y = 0$ at the position of the unstretched spring. This is more simply written as

$$m\,y'' + c\,y' + s\,y = -m\,g$$

HINT 36.8. *The Static Solution*
Find a simple formula for a constant solution $y_S = ?$ to the above differential equation forced by gravity. (Hint: Plug a constant into the left-hand side.) Explain the formula for your solution in terms of the strength of the spring and weight of the object. (Which parameter measures spring strength?)

What does the word "static" mean in this context? What is the derivative of your solution?

Recall that Theorem 23.7 of the main text says in particular that every solution of a homogeneous or autonomous second-order constant coefficient equation may be written as a linear combination of certain special exponential solutions. That theorem is a computational procedure, but it guarantees that you find the one and only solution and you find it by solving for unknown constants in a certain form.

HINT 36.9. *Gravity*
Suppose that $y[t]$ satisfies $my'' + cy' + sy = -mg$. Use simple reasoning and results from the course (which you cite) to show that we may write the solution in the form $y[t] = y_S + k_1 x_1[t] + k_2 x_2[t]$, for basic solutions $x_1[t]$ and $x_2[t]$ of $my'' + cy' + sy = 0$. (Hint: What equation does $y[t] - y_S$ satisfy?)

Conversely, if we let $y[t] = y_S + k_1 x_1[t] + k_2 x_2[t]$, for basic solutions $x_1[t]$ and $x_2[t]$ of $my'' + cy' + sy = 0$, then show that $y[t]$ satisfies $my'' + cy' + sy = -mg$.

Use the formula $y[t] = y_S + k_1 x_1[t] + k_2 x_2[t]$ to find a solution to

$$y(0) = 1$$
$$y'(0) = 1$$
$$my'' + cy' + sy = -mg$$

in the three cases $m = 1$, $s = 1$, $c = 3, 2, 1$.

While you are solving this, think about how you would program the computer to do this for you and, more generally, how to program it to solve forced initial value problems. You should include a program to solve these problems. The commands in the electronic assignment Exercise 23.7.2 of the main text need to be modified.

The physical meaning of the previous exercise is that basic homogeneous solutions permit us to adjust initial conditions without disturbing the applied force. The damping term, however, dissipates those "transient" solutions. Show this.

HINT 36.10. *Transients Die*
Show that every solution $y[t]$ of

$$my'' + cy' + sy = -mg$$

satisfies

$$\lim_{t \to \infty} y[t] = y_S \qquad \text{the constant steady-state solution}$$

by applying the exercise above and theory from the course.

36.5. Equations Forced by Sinusoids

Car suspensions are interesting only when we wiggle them. As a test case, imagine an applied force that is a perfect sinusoid,

$$my'' + cy' + sy = f_c \operatorname{Cos}[\omega t] + f_s \operatorname{Sin}[\omega t] \qquad \text{for constants } f_c, f_s$$

We want you to find a solution by guessing a form and substituting it into the equation. Assume that the solution has the form

$$y_S[t] = h \operatorname{Cos}[\omega t] + k \operatorname{Sin}[\omega t]$$

plug $y_S[t]$ into the left hand side of the equation, tidy up the algebra, and equate coefficients on the sines and cosines. You get the system of equations:

$$\begin{bmatrix} s - mw^2 & cw \\ -cw & s - mw^2 \end{bmatrix} \begin{bmatrix} h \\ k \end{bmatrix} = \begin{bmatrix} f_c \\ f_s \end{bmatrix}$$

You can solve this with Mathematica's linear solve or use the 2 by 2 Cramer's rule:

$$h = \frac{\det \begin{vmatrix} f_c & cw \\ f_s & s - mw^2 \end{vmatrix}}{\det \begin{vmatrix} s - mw^2 & cw \\ -cw & s - mw^2 \end{vmatrix}}$$

and the similar formula for k. We want to find an interpretation of this solution in terms of the limiting behavior of the system. We will call the solution $y_S[t] = h\,\text{Cos}[w\,t] + k\,\text{Sin}[w\,t]$ the "steady-state" solution for these particular values of h and k. We do not as yet know the actual solution that satisfies the initial conditions, but we do know Theorem 23.7 of the main text that tells us a form in which homogeneous solutions can be written.

HINT 36.11. *Single Frequency Forcing*
Suppose that $y[t]$ satisfies

$$m\,y'' + c\,y' + s\,y = f_c\,\text{Cos}[w\,t] + f_s\,\text{Sin}[w\,t]$$

Use simple reasoning and results from the course (which you cite) to show that we may write the solution in the form $y[t] = y_S[t] + k_1\,x_1[t] + k_2\,x_2[t]$, for basic solutions $x_1[t]$ and $x_2[t]$ of

$$m\,y'' + c\,y' + s\,y = 0$$

(Hint: What equation does $y[t] - y_S[t]$ satisfy?)
Conversely, if we let $y[t] = y_S[t] + k_1\,x_1[t] + k_2\,x_2[t]$, for basic solutions $x_1[t]$ and $x_2[t]$ of

$$m\,y'' + c\,y' + s\,y = 0$$

then show that $y[t]$ satisfies

$$m\,y'' + c\,y' + s\,y = f_c\,\text{Cos}[w\,t] + f_s\,\text{Sin}[w\,t]$$

As in the gravity solutions, the physical meaning of the previous exercise is that basic homogeneous solutions permit us to adjust initial conditions without disturbing the applied force. That is all the homogeneous solutions do:

HINT 36.12. *Transients Die*
Show that every solution $y[t]$ of $m\,y'' + c\,y' + s\,y = f_c\,\text{Cos}[w\,t] + f_s\,\text{Sin}[w\,t]$ satisfies

" $\lim_{t \to \infty} y[t] = y_S[t]$ "   the sinusoidal steady-state solution

by applying the exercise above and theory from the course. What is the meaning of the limit? Notice that $t \to \infty$ on the left-hand side, but not on the right. How can we formulate this correctly?

36.6. Nonhomogeneous IVPs

Use the formula $y[t] = y_S[t] + k_1 x_1[t] + k_2 x_2[t]$ to find a solution to

$$y(0) = 1$$
$$y'(0) = 1$$
$$m y'' + c y' + s y = f_c \cos[\omega t] + f_s \sin[\omega t]$$

in the three cases $m = 1$, $s = 1$, $c = 3, 2, 1$. While you are solving this, think about how you would program the computer to do this for you. Extend your old e-homework program for homogeneous problems to solve the full problem (see the discussion following Hint 36.9 above):

$$m y'' + c y' + s y = f_c \cos[\omega t] + f_s \sin[\omega t]$$
$$y(0) = y_0$$
$$y'(0) = y_1$$

Combine the homogeneous and the steady-state solutions,

$$y[t] = y_S[t] + k_1 y_1[t] + k_2 y_2[t]$$

set $t = 0$ in your program and solve for k_1 and k_2 just as you did in your program for the exact Mathematica solution of the homogeneous problem, but now take the initial value of $y_S(0)$ into account.

PROJECT 37

Resonance – Maximal Response to Forcing

Resonance is a peak in vibration as you vary the frequency at which you "shake" a system. For example, many old cars have speeds where they hum loudly, say at 47 mph. Slower, it's quieter and faster it's also quieter. We will see this phenomenon in the solution of the linear system

$$my'' + cy' + sy = \text{Sin}[\omega\, t]$$
$$y(0) = y_0$$
$$y'(0) = y_1$$

where we vary the frequency w and observe the amplitude of the response.

By our study of "transients" in Project 36, we know that the initial conditions do not matter for long-term behavior of the solutions. We are interested only in the amplitude of the steady-state solution.

The values of h and k in the steady-state solution,

$$y_S[t] = h\,\text{Cos}[\omega\, t] + k\,\text{Sin}[\omega\, t]$$

were found in Hint 36.11. You should verify the solutions:

$$h = \frac{-cw}{(cw)^2 + (s - mw^2)^2} \quad \text{and} \quad k = \frac{s - mw^2}{(cw)^2 + (s - mw^2)^2}$$

37.1. Some Useful Trig

Given h and k, we want to compute a and ϕ so that for all θ,

$$h\,\text{Cos}[\theta] + k\,\text{Sin}[\theta] = a\,\text{Sin}[\theta + \phi]$$

Conversely, given a and ϕ, we want to compute the values of h and k that make the equation hold for all θ. The number a is called the amplitude and the "angle" ϕ is called the "phase." (This is not the phase variable trick, but a different "phase.")

It is remarkable that any linear combination of sine and cosine is actually another sine. And conversely, that any phase shift may be written as a combination of only plain sine and cosine. The usefulness of this form comes up in the study of resonance below.

HINTS IN THE CASE WHERE h AND k ARE GIVEN:

Approach 1) We could use calculus to maximize $h\cos[\theta] + k\sin[\theta]$ for all values of θ. The max of $a\sin[\theta + \phi]$ is clearly a at the angle where the sine peaks.

Approach 2) For any pair of numbers (α, β) satisfying $\alpha^2 + \beta^2 = 1$ there is an angle ϕ, so that
$$(\alpha, \beta) = (\sin[\phi], \cos[\phi])$$

Why is this so? (Think of an appropriate parametric curve. What is the definition of radian measure? What does this problem have to do with the unit circle? Draw the vector (α, β). What is its length?)

If we know how to find ϕ so a given $(\alpha, \beta) = (\cos[\phi], \sin[\phi])$, let $a = \sqrt{h^2 + k^2}$, so that
$$(\alpha, \beta) = \left(\frac{h}{a}, \frac{k}{a}\right)$$

satisfies $\alpha^2 + \beta^2 = 1$ and produces an angle ϕ as above.

Use this angle to write
$$h\cos[\theta] + k\sin[\theta] = a\left(\frac{h}{a}\cos[\theta] + \frac{k}{a}\sin[\theta]\right)$$
$$= a\left(\sin[\phi]\cos[\theta] + \cos[\phi]\sin[\theta]\right)$$

Finish proving that these values of a and ϕ make
$$h\cos[\theta] + k\sin[\theta] = a\sin[\theta + \phi]$$

by using the addition formula for sine.

We won't need the converse part, but you can probably see why it holds once you find the formulas for a and ϕ in terms of h and k.

37.2. Resonance in Forced Linear Oscillators

HINT 37.1. RESONANCE QUESTION 1
How much is $a = a[\omega]$ when we write the solution
$$y_S[t] = h\cos[\omega t] + k\sin[\omega t]$$
in the form
$$y_S[t] = a[\omega]\sin[\omega t + \phi]$$

Express your answer in terms of the physical parameters m, c, and s as well as the variable forcing frequency w.
$$a[\omega] = ???$$

HINT 37.2. RESONANCE QUESTION 2
Simplify the expression for $a[\omega]$ and show that $a[\omega]$ is maximized when the square of its reciprocal is minimized. What frequency ω_M minimizes
$$b[\omega] = (s - m\omega^2)^2 + (c\omega)^2$$

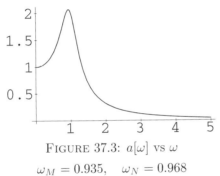

FIGURE 37.3: $a[\omega]$ vs ω

$$\omega_M = 0.935, \quad \omega_N = 0.968$$

This is called the resonant frequency, and now you should analyze it in various cases of the physical constants. Use the computer to plot $a[\omega]$ versus ω for various choices of the parameters m, c, and s. First observe that we have resonance only when $2\,c^2 < 4\,m\,s$. Next observe that if $2\,c^2 < 4\,m\,s$, then we must have $c^2 - 4\,m\,s < 0$. The autonomous system oscillates with "natural frequency"

$$\omega_N = \frac{\sqrt{4\,m\,s - c^2}}{2\,m}$$

provided that $c^2 - 4\,m\,s < 0$. Explain in your project what this autonomous oscillation means. Why is this called the "natural frequency?"

What happens when $2\,c^2 \geq 4\,m\,s$, but $c^2 < 4\,m\,s$? What happens if $c^2 > 4\,m\,s$? Verify this mathematical split with some computer plots or experiments.

37.3. An Electrical Circuit Experiment

This is an optional suggestion for an experiment to verify your theoretical calculations. The series R-L-C electrical experiments work very well, because the components are very nearly linear at low current. If you can get access to a variable resistor box, a variable capacitor box, an inductor (the bigger the inductance, the better), a wave generator, and an oscilloscope, all of which are in most engineering labs, you can easily build a lab apparatus that illustrates ALL of the things in the above project. An R-L-C circuit satisfies a linear second-order equation and the wave generator provides the forcing term. By varying the resistor and capacitor, you can see all the things we have learned – the graphs appear on the oscilloscope! (It is much harder to perform the mechanical oscillator experiments, because it is difficult to vary spring constants and damping friction.)

37.4. Nonlinear Damping

The most suspect term in our mechanical oscillator is the damping force

$$-c\,\frac{dy}{dt}$$

For damping caused by air resistance, we have investigated nonlinear laws like

$$-c\left[\frac{dy}{dt}\right]^{7/5}$$

(in the bungee project.) The nonlinear 7/5 power makes the damping force grow faster as the speed increases. This optional section asks how we might detect nonlinearity from the behavior of solutions of the system.

The frequency of oscillation of a linear system

$$m\frac{d^2y}{dt^2} + c\frac{dy}{dt} + sy = 0$$

is always the same (provided it oscillates). What is this frequency?

Run the computer program **ComplexRoots** and observe that initial values that begin on a line in the phase plane remain in line. Why is this just the geometric representation of the fact that the frequency is independent of the initial condition?

Conduct some computer experiments with the system

$$m\frac{d^2y}{dt^2} + c\left[\frac{dy}{dt}\right]^{7/5} + sy = 0$$

beginning with the **ComplexRoots** program, by modifying it to solve the nonlinear equation. With suitable choices of the parameters and suitable scales for your flow, you will be able to see nonlinearity arising as frequencies that depend on the initial conditions. Explain what you observe. If the nonlinearity of a shock absorber is small, say

$$c\left[\frac{dy}{dt}\right]^p$$

with $p \approx 1$, will this be easy to detect in a physical system? What if p is large, say $p = 3$?

PROJECT 38

A Notch Filter – Minimal Response to Forcing

One can build an r-c circuit with two loops that has the property that there is a local minimum in its response to oscillatory forcing. This local minimum can be used to filter out that frequency.

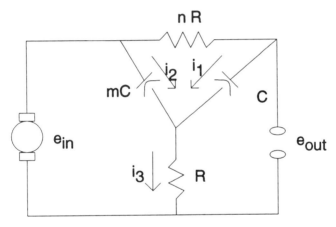

FIGURE 38.1: A Two-Loop r-c Circuit

We derive the equations of the system using Kirchhoff's laws for voltage and current, Ohm's law for resistors, and Coulomb's law for capacitors. Kirchhoff's voltage law says that the signed sum of voltages around a closed loop is zero. Kirckoff's current law says the signed sum of all currents into a node is zero.

The currents are shown in Figure 38.1. There is no current through the open output voltage location e_{out}. At the central node, i_1 and i_2 come in, while i_3 goes out, so we must have $i_3 = i_1 + i_2$, eliminating i_3.

38.1. The Laws of Kirchhoff, Ohm, and Coulomb

Ohm's law says that the voltage across a resistor of resistance r carrying a current i is $r \cdot i$, so the voltages across the resistors are

$$e_{nr} = n\,r\,i_1 \quad \text{and} \quad e_r = r\,i_3 = r\,(i_1 + i_2)$$

with the sign in the direction of the currents.

Coulomb's law says that the voltage across a capacitor of capacitance c with a charge q is $e_c = \frac{1}{c} q$ Charge is the integral of the current passing through it, so

$$e_c = \frac{1}{c} q_1 = \frac{1}{c} \int_0^t i_1[s] \, ds \quad \text{and} \quad e_{mc} = \frac{1}{mc} q_2 = \frac{1}{mc} \int_0^t i_2[s] \, ds$$

with the sign in the direction of the currents.

Current is a flow of charge, or charge is an accumulation of current. By the second half of the Fundamental Theorem of Calculus (see Chapter 12 of the core text),

$$\frac{dq_1}{dt} = \frac{d}{dt} \int_0^t i_1[s] \, ds = i_1[t] \quad \text{and} \quad \frac{dq_2}{dt} = i_2[t]$$

so we may express the resistor voltages as

$$e_{nr} = nr \frac{dq_1}{dt} \quad \text{and} \quad e_r = r i_3 = r \left(\frac{dq_1}{dt} + \frac{dq_2}{dt} \right)$$

The sum of the voltages around the upper loop makes

$$e_{nr} + e_c = e_{mc}$$
$$nr \frac{dq_1}{dt} + \frac{1}{c} q_1 = \frac{1}{mc} q_2$$
$$\frac{dq_1}{dt} = \frac{-1}{nrc} q_1 + \frac{1}{mnrc} q_2$$

The sum of voltages around the lower loop makes

$$e_{in} = e_{mc} + e_r$$
$$e_{in} = \frac{1}{mc} q_2 + r \left(\frac{dq_1}{dt} + \frac{dq_2}{dt} \right)$$
$$\frac{dq_2}{dt} = -\frac{dq_1}{dt} - \frac{1}{mrc} q_2 + \frac{1}{r} e_{in}[t]$$

We substitute the first equation above for $\frac{dq_1}{dt}$ on the right side to obtain the pair of equations,

$$\frac{dq_1}{dt} = \frac{-1}{nrc} q_1 + \frac{1}{mnrc} q_2$$
$$\frac{dq_2}{dt} = \frac{1}{nrc} q_1 - \left(\frac{1}{mnrc} + \frac{1}{mrc} \right) q_2 + \frac{1}{r} e_{in}[t]$$

This is a nonautonomous linear system of the form

$$\begin{bmatrix} \frac{dx}{dt} \\ \frac{dy}{dt} \end{bmatrix} = \begin{bmatrix} -a_x & a_y \\ b_x & -b_y \end{bmatrix} \begin{bmatrix} x \\ y \end{bmatrix} + \begin{bmatrix} 0 \\ f[t] \end{bmatrix}$$

where the forcing term is $f[t] = e_{in}[t]/r$, $a_x = \frac{1}{nrc} = b_x$, $a_y = \frac{1}{mnrc}$, and $b_y = \left(\frac{1}{mnrc} + \frac{1}{mrc} \right)$.

HINT 38.2. *Transients*
Use Theorem 24.3 of the core text to show that all solutions of

$$\frac{dq_1}{dt} = -a_x\, q_1 + a_y\, q_2$$
$$\frac{dq_2}{dt} = a_x\, q_1 - b_y\, q_2$$

tend to zero as $t \to \infty$, *when* a_x, a_y, *and* b_y *are as above.* (Note: $b_y > a_y$.)

We are interested in the way the steady-state solution for $e_{out}[t]$ depends on ω in the case where $f[t] = \text{Sin}[\omega t]/r$. Applications of Kirchhoff's voltage law around different loops give two different expressions for the output voltage:

$$e_{out}[t] = e_{in}[t] + \frac{1}{c} q_1[t] - \frac{1}{mc} q_2[t]$$
$$e_{out}[t] = e_{in}[t] - n\, r\, i_1$$

38.2. Steady-State Solution

Now we turn our attention to the steady-state solution when $e_{in}[t] = \text{Sin}[\omega t]$, or $f[t] = \text{Sin}[\omega t]/r$.

HINT 38.3. *Steady State Solution*
We will assume that the solution of the forced equations

$$\frac{dq_1}{dt} = -a_x\, q_1 + a_y\, q_2$$
$$\frac{dq_2}{dt} = a_x\, q_1 - b_y\, q_2 + \text{Sin}[\omega t]/r$$

may be written in the form

$$q_1[t] = h_1\, \text{Cos}[\omega t] + k_1\, \text{Sin}[\omega t]$$
$$q_2[t] = h_2\, \text{Cos}[\omega t] + k_2\, \text{Sin}[\omega t]$$

By substituting these forms into the equations, we will be able to solve for the constants h_1, h_2, k_1, k_2 *and verify our assumption.*

(1) *Substitute* q_1 *and* q_2 *into the first equation and show that the coefficients satisfy:*

$\text{Cos}[\omega t]$	$\text{Sin}[\omega t]$
$\omega\, k_1 = -a_x\, h_1 + a_y\, h_2$	$-\omega\, h_1 = -a_x\, k_1 + a_y\, k_2$

(2) *Substitute* q_1 *and* q_2 *into the second equation and show that the coefficients satisfy:*

$\text{Cos}[\omega t]$	$\text{Sin}[\omega t]$
$\omega\, k_2 = a_x\, h_1 - b_y\, h_2$	$-\omega\, h_2 = a_x\, k_1 - b_y\, k_2 + 1$

(3) *Write the coefficient equations from parts (1) and (2) in the form:*

$$\begin{aligned}
-a_x\, h_1 + a_y\, h_2 - \omega\, k_1 + 0\, k_2 &= 0 \\
\omega\, h_1 + 0\, h_2 - a_x\, k_1 + a_y\, k_2 &= 0 \\
a_x\, h_1 - b_y\, h_2 + 0\, k_1 - \omega\, k_2 &= 0 \\
0\, h_1 + \omega\, h_2 + a_x\, k_1 - b_y\, k_2 &= -1/r
\end{aligned}$$

(4) *For example, you could use Mathematica to solve these equations as follows:*

aX = 1/(n r c)
aY = 1/(m n r c)
bX = 1/(n r c)
bY = (1/(m n r c) + 1/(m r c))
mat = {{-aX, aY,-w,0},
　　　　{w,0,-aX,aY},
　　　　{bX,-bY,0,-w},
　　　　{0,w,bX,-bY}}
rhs = {0,0,0,-1/r}
{h1,h2,k1,k2} = **LinearSolve[mat,rhs]**

We know from above that

$$e_{out}[t] = e_{in}[t] + \frac{1}{c} q_1[t] - \frac{1}{mc} q_2[t]$$

$$= \sin[\omega\, t] + \frac{1}{c}(h_1\, \cos[\omega\, t] + k_1\, \sin[\omega\, t]) - \frac{1}{mc}(h_2\, \cos[\omega\, t] + k_2\, \sin[\omega\, t])$$

HINT 38.4. *Show that*

$$e_{out}[t] = h_3\, \cos[\omega\, t] + k_3\, \sin[\omega\, t]$$

where $h_3 = \frac{h_1}{c} - \frac{h_2}{mc}$ and $k_3 = \frac{k_1}{c} - \frac{k_2}{mc} + 1$.

Now we may use the trig trick of Hint 37.1 to write

$$e_{out}[t] = a[\omega]\, \sin[\omega\, t + \phi[\omega]]$$

where $a[\omega] = \sqrt{h_3^2 + k_3^2}$.

HINT 38.5. *Continue your computer program, defining the square of the output amplitude,* $(a[\omega])^2$:

h3 = **Simplify[h1/c - h2/(m c)]**
k3 = **Simplify[k1/c - k2/(m c) + 1]**
a2 = **Simplify[(h3)^2 + (k3)^2]**

38.3. A Check on $a[t]^2$

We can also use

$$e_{out}[t] = e_{in}[t] - n\,r\,i_1 = e_{in} - n\,r\,\frac{dq_1}{dt}$$
$$= \text{Sin}[\omega\,t] - n\,r\,(-\omega\,h_1\,\text{Sin}[\omega\,t] + \omega\,k_1\,\text{Cos}[\omega\,t])$$

HINT 38.6. *Show that*

$$e_{out}[t] = h_4\,\text{Cos}[\omega\,t] + k_4\,\text{Sin}[\omega\,t]$$

where $h_4 = -n\,r\,\omega\,k_1$ and $k_4 = 1 + n\,r\,\omega\,h_1$.
Show that we should also have

$$a[\omega]^2 = \left((n\,r\,\omega\,k_1)^2 + (1 + n\,r\,\omega\,h_1)^2\right)$$

and verify that this agrees with your other expression for $a^2[\omega]$ with help from the computer.

38.4. Where's the Min?

Recall that we seek a minimum of the output amplitude, $a[\omega]$, because we want to design the circuit to minimize its response to certain frequencies, ω. How do we minimize a function $b[\omega] = a^2[\omega]$ defined for $0 < \omega < \infty$?

Although the output expression from the computer looks pretty complicated, we know that we need to look for critical points in order to minimize. We also need some additional information, since we do not have a compact interval minimization problem. The computer can differentiate for us.

HINT 38.7. *Use the computer to differentiate $a[\omega]^2$,*

da = Simplify[D[a2,w]]

HINT 38.8. *Show that the numerator of $\frac{d(a^2)}{d\omega}$ is a positive multiple of*

$$(m\,n\,c^2\,r^2\,\omega^2)^2 - 1$$

HINT 38.9. *Prove that $a[\omega]$ has a minimum at $\omega_0 = 1/(r\,c\,\sqrt{m\,n})$. Be complete in your reasoning. Why can't the minimum occur between 0 and ω_0? Why can't the minimum occur between ω_0 and ∞?*
What is the maximum of $a[\omega]$?

Round out your project with some specific choices of the parameters and a plot of the amplitude as a function of frequency.

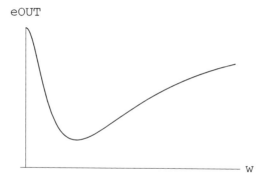
FIGURE 38.10: A Minimal Response to Forcing

Chapter 13

Applications in Ecology

Birth Rate, Carrying Capacity, and Hunting: The first project in this part reviews the ideas of per capita birth rate and carrying capacity, then explores two kinds of hunting when Voodoo the barn cat discovers the prolific mouse population in the back field.

A Journey to the Sogapalag Islands: The next two projects concern continuous mathematical ecology of interacting species. They are somewhat simplified, so we have devised a story about Prof. Debrainman's discovery of the Sogapalag Islands to help you understand the simplifications underlying these projects.

Over a century ago, Charles Darwin's journey resulted in a theory that caused a revolution in biology. Now, thanks to the efforts of our own Professor Debrainman, we have available to us data from the newly discovered Sogapalag Islands. While Prof. Debrainman is a great environmental biologist, his math skills are lacking, rendering him incapable of fully utilizing the data available to him. He has come to us for help in organizing and understanding his observations.

According to the professor, Sogapalag is unique in that it has a very restricted food web, which makes it ideal for a study of interactions. On the lowest trophic level, we have the grass and bushes with dense foliage, which are indigenous to all the islands. The second trophic level consists of a single species of rabbits, not unlike those we have in America, and an indigenous species of sparrows. The birds can travel to all the islands, but avoid those without many bushes. The rabbits are confined to only some of the islands. The two species share common resources and so must compete against one another, as well, when they try to inhabit the same island. This competition is fiercest when the bushes are sparse, because then they must both live on the native grass.

The third species on the highest trophic level is an indigenous weasel-like animal unlike any ever seen before. The average specimen stands a foot tall, and both males and females have long fangs that are clearly visible from a great distance. The weasels are ferocious predators that prey on the rabbits and finches alike, being excellent leapers. Like the land bound rabbits, the weasels inhabit only some of the islands.

These three trophic levels with different mixes on different islands allow for a diverse array of interactions with one another. Fortunately, the various islands have different combinations of species, so we can begin with simple interactions before we move on to the study of the most complicated islands. We want to study these interactions, but this study introduces another element. Professor Debrainman has brought a cadre of voracious graduate students with him.

In general, the professor wants to know the mathematical implications of the different interactions in terms of the long-term behavior of the different populations' dynamics. Unfortunately, the professor not only cannot perform mathematical computations, he also cannot comprehend those already done. You must therefore be careful to explain all calculations and results in terms of their biological implications. Any situation which would harm the ecosystem must be identified – it's hard to win a Nobel Prize for studying animals that became extinct midway through your study – especially if the extinction is a result of your interaction with the species.

A Sustainable Harvest Level for Whales: The third project examines a proposal made to the international whaling commission concerning a sustainable harvest level for a certain species of whale.

PROJECT 39

Logistic Growth with Hunting

A field of logistically growing mice is visited by Voodoo the barn cat. This project begins with a review of some of the basic ideas of Chapters 8 and 21, especially Examples 21.3 — 5 and Problem 21.4.

39.1. Basic Fertility

Each spring, mice reproduce prolifically in the field behind my house. One Monday before work, I counted about 1000 mouse pairs per acre. When I came home the mouse census had increased to 1100 per acre.

VARIABLES
$$t = \text{time in days}$$
$$x = \text{mouse couples per acre}$$

PARAMETERS
We begin with the single parameter r, the per capita birth rate of mice.

HINT 39.1. *Explain why the differential equation*
$$\frac{dx}{dt} = r\,x$$
says that the instantaneous per capita birth rate of mice is r. What are the units of r? What does "per capita" mean? Explain why more mice are born when x is large than when x is small, even though r is constant.

HINT 39.2. *Show that the solution to the initial value problem*
$$x[0] = 1000$$
$$\frac{dx}{dt} = r\,x$$
is $x[t] = 1000\,e^{r\,t}$. (Hint: See text Section 8.2.)

HINT 39.3. *Use the information $x = 1100$ in 8 hours to show that $r \approx 0.286$. How often does each mouse couple have a pair of babies?*

When the mouse population is small compared with the food and shelter available to them in my small 7-acre field, the birth rate can continue at the per capita rate $r \approx 0.286$. However, this cannot continue for even 1 month.

HINT 39.4. *Estimate how long growth at the rate $\frac{dx}{dt} = rx$ could continue until the back field is piled 3 deep in mice over its entire 7-acre area. (A mouse is about 3 inches long and 1 inch wide. One acre is 43,560 square feet. See the **ExpGth** program from Chapter 28 of the main text.)*

39.2. Logistic Growth

ANOTHER PARAMETER: c

We introduce a parameter c called the carrying capacity of the ecosystem. Our introduction is mathematical, and your job is to explain the biological significance of this parameter.

The former owner of my field said he noticed that the birth rate of the mice dropped off sharply each spring as the population density reached about 5000 mouse couples per acre. When the population is small, the basic fertility of mice is the constant r,

$$\frac{1}{x}\frac{dx}{dt} = r$$

but as x grows toward 5000, food and shelter become difficult for the mice and the per capita growth declines toward zero.

HINT 39.5. *The Logistic Growth Law*
Explain how the differential equation

$$\frac{1}{x}\frac{dx}{dt} = r\left(1 - \frac{x}{c}\right)$$

says that the per capita rate of growth of mice is a decreasing function with a basic fertility rate of r but a limiting population of c mouse couples per acre. Show that the limit of x as t tends to infinity is c,

$$\lim_{t \to \infty} x[t] = c$$

(Hint: See Problem 21.4 of the text.) Notice the contrast of this logistic growth law with Hint 39.4

HINT 39.6. *Use the value $c = 5000$ and return to the data that 1000 mice became 1100 in 8 hours to estimate the true value of r. Solve the differential equation with a selection of initial conditions using the FLOW1D program from Chapter 21 of the main text. Solve the differential equation exactly, as well, using the computer if you wish. How many mice are there 1 month after we observed 1000 per acre?*

39.3. Voodoo Discovers the Mice

Voodoo the barn cat came with our property. I believe he feels that HE owns the property and has to tolerate a new tenant. He lets us feed and pet him and seems to like the new insulated cat house I built. But hunting is Voodoo's life. We prefer that he leave the song birds alone, but when the mice start coming in the house, we're grateful for his diligence.

Voodoo has noticed all the mouse activity in the back field and has decided to concentrate his efforts there. Being well-fed and preferring rabbits anyway, he often brings us some of his extra catch, so we have some idea of his mousing success.

ANOTHER PARAMETER: h

We notice that Voodoo's catch increases with increasing density of mice, so we want to explore some descriptions of his impact on the mouse population. We introduce another parameter h - Voodoo's hunting success rate. Our first try will be hunting success that increases linearly with population.

HINT 39.7. *Linear Hunting*
Suppose Voodoo's hunting success rate is a linear function of mouse density,

$$\frac{dx}{dt} = r\,x\left(1 - \frac{x}{c}\right) - h\,x$$

Explain the biological meaning of the term $h\,x$. Show that

$$\lim_{t \to \infty} x[t] = c\left(1 - \frac{h}{r}\right)$$

And discuss the importance of the ratio h/r.

The previous hint assumes that Voodoo will be proportionately as successful at low densities as he is at high densities. Here is another possible model of hunting success:

HINT 39.8. *Nonlinear Hunting*
Suppose the mouse population is affected by Voodoo as follows,

$$\frac{dx}{dt} = r\,x\left(1 - \frac{x}{c}\right) - h\,x^2$$

Show that

$$\lim_{t \to \infty} x[t] = \frac{c}{1 + h/r}$$

Notice that the linear hunting model predicts that Voodoo can hunt the mice to extinction, whereas the nonlinear one does not. Why is this? In particular, what is the difference between the hunting effects $h\,x$ and $h\,x^2$ – especially as the mouse population gets low?

PROJECT 40

Predator-Prey Interactions

The professor is very eager to begin a complete physiological and anatomical characterization of the species on Sogapalag. Next week, a group of biology graduate students from his lab is scheduled to invade the island he named "Bunny" in search of a variety of specimens of all indigenous species. Bunny Island has only rabbits and weasels, who up to now have managed to coexist, though somewhat precariously since the windward weather is tough on grass and almost impossible for bushes.

The students are hungry for results to apply to their theses and the professor is eager to encourage them in his quest for a Nobel Prize. The specimens will be taken back to the United States, where they can be dissected for a complete characterization of their anatomy. This will allow a comparison of their internal features with those of traditional sparrows, rabbits, and weasels. The graduate students plan to capture as many of each species as possible, and so can be thought to exert a "fishing effect" on the population. In defining such a relationship, we must first characterize the variables we will be using. We are hoping your model can help us set limits on the amount of collection that we permit the graduate students to do.

40.1. Bunny Island

The dynamics of the interactions on Bunny Island are described by the following differential equations.

VARIABLES:

(1) t = time in weeks
(2) x = density of rabbit pairs per square mile
(3) y = density of weasel pairs per square mile

PARAMETERS:

(1) a = the intrinsic per capita growth rate of the rabbits (1/time)
(2) b = the predator's effectiveness in terms of prey capture
(3) c = the predator's intrinsic death rate (1/time)
(4) e = the predator's efficiency in terms of prey consumption
(5) f = the gathering effectiveness of grad students, per unit of specimen

$$\frac{dx}{dt} = ax - bxy - fx = a\,x((1-f/a) - b\,y/a)$$
$$\frac{dy}{dt} = -cy + exy - fy = -c\,y((1+f/c) - e\,x/c)$$

HINT 40.1. *A Model of Bunnies, Weasels, and Grad Students*
Your first task is to describe what these equations say about the interactions. Remember that poor Professor Debrainman needs clear biological descriptions. Try to be as biological as you can. We offer a start as follows.

When there is no "fishing" by the graduate students, $f = 0$, our model is the Lotka Volterra model studied in Chapter 22 of the text. The parameter a represents the intrinsic fertility of the rabbits. When $y = 0$ and x is small, the growth of rabbits is

$$\frac{dx}{dt} \approx a\,x \qquad \text{or} \qquad \frac{1}{x}\frac{dx}{dt} \approx a$$

which says that each pair of rabbits has a babies per week. The units are "new bunny pairs per bunny pair per week." Canceling units of bunnies, we obtain a in units of "1/weeks."

The term $-b\,x\,y$ represents the hunting effectiveness of the weasels. Let us be more explicit. The units of the whole term $-b\,x\,y$ are "(negative or captured) bunnies per unit time." If there are lots of weasels, y is big, then lots of bunnies get caught, but if there aren't many bunnies, x is small, then the hunting is tougher and fewer bunnies get caught. In other words, the whole term takes the density of rabbits and the density of weasels both into account. The units of the whole term are "bunnies per week = units of parameter b × bunnies × weasels," so the units of b are "bunnies (caught) per week per bunny per weasel = 1/(weasel weeks)."

Your project should explain all the terms to Prof. Debrainman and give the units as well. In particular, explain to the professor that the success of his students depends on the densities of the two species that exist on the island. If these are low, the students will have trouble collecting samples.

The professor is rather worried about the effect of this "fishing" on his isolated little ecosystem. By manipulating the mathematical model for this relationship, we can predict what effect the graduate students will have on the rabbit and weasel populations.

HINT 40.2. *Mathematical Experiments*
*Continue your project with some computer based on the **Flow2D** program or the Fox-Rabbit example program. We want to understand the effect of increased "fishing" on the fragile ecosystem of Bunny Island, but first you must understand the basic model. Begin your experiments with $a = 0.1$, $b = 0.005$, $c = 0.04$, and $e = 0.00004$ and no graduate students, $f = 0.0$.*
Here are some basic questions about the model with these parameter values:

(1) *How often does a pair of rabbits have babies if they typically give birth to four male and four female babies?*
(2) *What are the direct interpretations of the other parameters?*

(3) You will see that there is an equilibrium of $x_e = 1000$ and $y_e = 20$ for the undisturbed island. Try various initial conditions, so that you can understand the nature of this equilibrium. For example, what happens to the populations if there is a displacement from equilibrium, say, that the initial number of weasels is 10, 20, 30, 40 and the initial number of rabbits is 1000 in each case? A flow picture is shown in Figure 40.3.

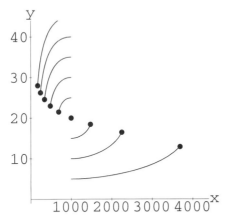

FIGURE 40.3: Rabbits and Weasels

Now introduce a small amount of "fishing," $f = 0.05$, and recompute your computer flow. What happens to the flow dynamics? What happens to the point of equilibrium?

Increase the fishing to $f = 0.1$ and compute the flow. What does the flow look like in this case? Do the weasels survive? Do the rabbits?

Let the graduate students have full rein, $f = 0.5$. Does anything survive?

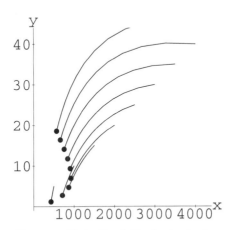

FIGURE 40.4: Grad Students Arrive

HINT 40.5. *Mathematical Conjectures*
Analyze the results of your mathematical experiments and describe the effect of various amounts of fishing on this environment. We observed three rather different types of dynamics in our experiments. Experimentally, what happens as f increases?

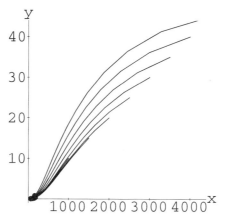

FIGURE 40.6: Overfishing

HINT 40.7. *Mathematical Proofs*
Use the theory of Chapter 22 to formulate a mathematical explanation for the various effects of increasing the fishing f from 0 to higher and higher values. Plot the curve of the equilibrium point (x_e, y_e) by hand parametrically depending on the parameter f. Remember that a parametric line will have the form

$$\begin{bmatrix} x_e \\ y_e \end{bmatrix} = \mathbf{P} + f\mathbf{D}$$

when f is the parameter. Give general symbolic answers in terms of a, b, c, e, and f, so that the professor can use the results on different islands where the natural parameters may be different. Hint: Use the computer to help where convenient, for example,
Solve$[\{ax - by - fx == 0, -cy + ex - fy == 0\}, \{x, y\}]$
What is the limit you must put on f in terms of the other parameters in order that the graduate students don't destroy one or both species?

The terms $-fx$ and $-fy$ represent collection of species all the time. Suppose that the graduate students only visit the island seasonally. One rough approximation for their hunting of samples might be a sinusoidal "fishing" term like

$$F[t] = f \cdot (\text{Sin}[p\,t + q] + 1)/2$$

so that the captured rabbits and weasels are given by $F[t] \cdot x$ and $F[t] \cdot y$, yielding the equations

$$\frac{dx}{dt} = ax - bxy - xF[t]$$
$$\frac{dy}{dt} = -cy + exy - yF[t]$$

These equations are non-autonomous, so the direction field flow analysis is not appropriate. However, explicit solutions can be used to make some experiments.

HINT 40.8. *Seasonal Collection*
Modify the **AccDEsol** program to test the effect of seasonal visits by graduate students collecting samples. In the constant sampling case, you found a threshold value for the constant

f where the natural dynamics of Bunny Island changed qualitatively. Use this value for f in the seasonal sampling, and assign values of p and q so that $F[0] = 0$, no students are on the island at midnight of a new year, and the oscillation of $F[t]$ is completed each year. When are the most graduate students visiting Bunny Island? What is their highest rate of collection? Be sure to run your model for several years. (Recall that t is in weeks.)

Does the seasonal relief from sampling allow the ecology on Bunny Island to recover? What happens if the students visit part of each semester, instead of annually.

Bunny Island is enormous, with vast expanses of grass, seemingly with no limit to the number of rabbits it can support, but neighboring Rabbit Island is much rockier. There still are few bushes and no sparrows, but now there is a limit to the total number of rabbits it can support.

40.2. Rabbit Island

Add a carrying capacity term to the rabbit-weasel equations:

$$\frac{dx}{dt} = ax - ax^2/k - bxy - fx = a\,x((1 - f/a) - x/k - b\,y/a)$$
$$\frac{dy}{dt} = -cy + exy - fy \qquad = -c\,y((1 + f/c) - e\,x/c)$$

and repeat the steps in the analysis of the ecosystem of Bunny Island. We observed a somewhat surprising added stability in this model. It actually seemed better able to cope with the seasonal onslaught of graduate students.

HINT 40.9. *The Model*
Explain the new term $-a\,x^2/k$ to Professor Debrainman in biological terms. For example, tell him the dynamics in the case where there are no weasels. (You might want to review the logistic equation from Chapter 22.) Give units and so forth. Why is k called the carrying capacity? How might Debrainman measure this?

HINT 40.10. *Basic Experiments*
Continue your project with some computer experiments based on the **Flow2D** program or the Fox-Rabbit example program. We want to understand the effect of increased "fishing" on the ecosystem of Rabbit Island, but first you must understand the basic model. Begin your experiments with $a = 0.1$, $b = 0.005$, $c = 0.04$, and $e = 0.00004$, no graduate students, $f = 0.0$, and a large carrying capacity, $k = 5000$.

Gradually decrease the carrying capacity k to see how that affects the dynamics.
Finally, introduce the graduate student collection of specimens, $f > 0$.

HINT 40.11. *Observations and Conjectures*
Make some biological observations about the behavior of your new model. For example, we felt that the new island was somewhat more stable in a sense. Even after a disturbance, when left alone, things returned to equilibrium in time, at least when k was large and f was small.

How small could we make k? How large could we make f?

HINT 40.12. *Proofs*

Formulate mathematical criteria for where the equilibrium occurs in terms of the parameters. Use the local stability theorem of Chapter 24 of the main text to compare the stability types of the equilibria of Bunny Island with those of Rabbit Island.

What is the mathematical limit on smallness of k and largeness of f after which the ecology of Rabbit Island is altered?

HINT 40.13. *Seasonal Collection*

*Modify the **AccDEsol** program to test the effect of seasonal visits by graduate students collecting samples on Rabbit Island. Be sure to watch for several years. We noticed a surprising difference between Bunny Island and Rabbit Island when we sampled seasonally with a peak near the threshold of sampling. Rabbit Island seemed better able to recover.*

PROJECT 41

Competition and Cooperation between Species

The concept of an interspecific relationship is not new – in fact, we have been dealing with one such relationship all along – a predator-prey interaction is an interspecific relationship. However, there are other interspecific relationships manifested on the Sogapalag island. While these relationships are fascinating to study as factors contributing to the predator prey interactions, the professor fears that the math involved in such a computation would be entirely over his head. Luckily, he finds several islands with sparrows and rabbits, but no weasels. We will therefore restrict our studies to include only two species. You might, however, try to discuss hypothetical implications of these relationships on rabbits, weasels, and sparrows for islands that have all three species.

41.1. Biological Niches

Biologically, we say that two species are competing when the presence of each detracts from the growth of the other. Rabbits and sparrows have partially overlapping niches, that is, they both eat grass and bushes, but the sparrows prefer bushes and the rabbits prefer grass. In general, a niche is defined as the collection of all necessary resources that define a species' way of life. Since the Sogapalag Islands have only grass and bushes, rabbits and sparrows, and weasels, the niches of rabbits and sparrows overlap by the extent to which they cannot choose their separate preferences in food and shelter. According to biologists, no two species may share the same niche, as one will inevitably be a superior competitor, driving the other to extinction or to develop a new niche. However, niches may partially overlap, with one or more common resources.

A mathematical model for competition between species can be formulated as the differential equations:

$$\frac{dx}{dt} = mx\left(1 - \frac{x}{h} - b\frac{y}{k}\right)$$
$$\frac{dy}{dt} = ny\left(1 - c\frac{x}{h} - \frac{y}{k}\right)$$

where
VARIABLES:

(1) t = time in weeks
(2) x = density of hundreds of rabbit pairs per square mile
(3) y = density of hundreds of sparrow pairs per square mile

We have already analyzed two cases of this model in Chapter 22 of the text. The first, titled "competition," was written in simplified form, but equivalent to

$$dx = m\,x\left(1 - \frac{x}{7} - \frac{5}{7}\frac{y}{5}\right) dt$$
$$dy = n\,y\left(1 - \frac{7}{10}\frac{x}{7} - \frac{y}{5}\right) dt$$

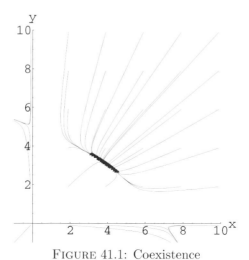

FIGURE 41.1: Coexistence

The second case of this model was titled "Fierce Competition" and was

$$dx = m\,x\left(1 - \frac{x}{10} - \frac{7}{5}\frac{y}{7}\right) dt$$
$$dy = n\,y\left(1 - \frac{10}{7}\frac{x}{10} - \frac{y}{7}\right) dt$$

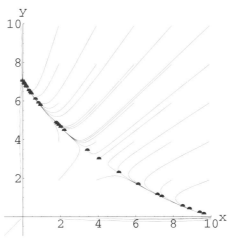

FIGURE 41.2: Weak Exclusion

We have purposely written the equations with the parameters b and c times the fractions $\frac{y}{k}$ and $\frac{x}{h}$. The appetites of rabbits and sparrows are different, corresponding to their different sizes, so we want to measure the effect of a number of sparrows on the rabbits in terms of the carrying capacity of sparrows. For example, without the other species, the environment might support 1000 sparrow couples or 400 rabbit couples per square mile. If there are 500 sparrow couples, half the environment's capacity, $b \times \frac{1}{2}$ tells us the negative affect on rabbits. All things being equal, you might expect "fair" competition to result in half rabbits and half sparrows, but a "fair" meaning to 'half and half' should take the various appetites into account. Half and half in terms of carrying capacity would mean 500 sparrow couples and 200 rabbit couples.

HINT 41.3. *The Intensity of Competition*
Begin your work on competing species by writing a biological explanation of the model for Professor Debrainman. The parameters h and k are the carrying capacities of an island for each species when the other is absent. Explain this to the professor by reviewing the logistic equation from Chapter 22.

The parameters m and n are the intrinsic fertilities of the species, similar to the a parameter in the predator - prey model above.

Explain the coefficients b and c carefully to the professor in terms of how effectively one species competes against another.

Give the units of each parameter with your explanation. And remember that the explanations are supposed to be as biological as possible, so that we don't terrorize the poor professor.

The parameters b and c are the most difficult to explain. You may want to think of some special cases to help. For example, if bushes are scarce so that both rabbits and sparrows rely mostly on the same resource, grass, then rabbits may heavily affect the sparrow population, who are rendered homeless if their bush is eaten. Sparrows may gobble grass that the bunnies can't then eat, but they will be at a disadvantage. What sorts of choices of b and c describe this situation?

A robust island with plenty of grass and bushes will allow the species to live in separate niches. The sparrows eat only a little grass and the rabbits eat only a few bushes. In this case, the impact of each species on the other is smaller. What are good choices of b and c to describe this situation?

You may want to come back to review your explanation after you have worked more on the mathematical experiments and analysis.

You should review the geometrical analysis of the two competition examples from Chapter 22 of the text. The first example leads to coexistence of rabbits and sparrows, but the second usually leads to the extinction of one or the other, but not both. This is "competitive exclusion," but a weak kind where the surviving species depends on the relative sizes of the two populations at the start of observation.

HINT 41.4. *Mathematical Experiments*
Use the computer program **Flow2D** or the **CowSheep** example to try some mathematical experiments with various values of the parameters. Begin with $m = n = 1$, $h = 7$, $k = 10$ and then run cases for the following islands.

(1) *Peaceful Island*, $b = 2/3$, $c = 3/4$.
(2) *Bushy Island*, $b = 3/2$, $c = 3/4$.
(3) *Grassy Island*, $b = 2/3$, $c = 4/3$.
(4) *Anxious Island*, $b = 3/2$, $c = 4/3$.

HINT 41.5. *Observations and Conjectures*
Summarize your mathematical experiments in the form of some basic conjectures about the two species competition model. What do you think the controlling mathematical factor is in determining the outcome of the competition?

Some of the algebra is simpler if we re-write the basic equations in terms of new variables,

$$u = \frac{x}{h} \quad \text{and} \quad v = \frac{y}{k}$$
$$\frac{du}{dt} = \frac{1}{h}\frac{dx}{dt} \quad \frac{dv}{dt} = \frac{1}{k}\frac{dy}{dt}$$

The competition equations become

$$\frac{1}{h}\frac{dx}{dt} = m\frac{1}{h}x\left(1 - \frac{x}{h} - b\frac{y}{k}\right) \quad \Leftrightarrow \quad \frac{du}{dt} = mu(1 - u - bv)$$
$$\frac{1}{k}\frac{dy}{dt} = n\frac{1}{k}y\left(1 - c\frac{x}{h} - \frac{y}{k}\right) \quad \Leftrightarrow \quad \frac{dv}{dt} = nv(1 - cu - v)$$

HINT 41.6. *Proofs:*
Prove your conjectures by making the compass heading direction fields similar to those in the text, but strictly in terms of the parameters h, k, b and c. Recall that the direction field is vertical, or $dx = 0$, on the line $1 = x/h + by/k$. This line crosses the x-axis at h and the y-axis at k/b. The direction field is horizontal, or $dy = 0$ on the line $1 = cx/h + y/k$. This crosses the y-axis at k and the x-axis at h/c. The relative locations of the pairs of points $(h, 0)$ and $(h/c, 0)$ and $(0, k)$ and $(0, k/b)$ corresponding to the four cases above can be filled in on the blank graphs in Figure 41.7 to get you started. (If you use the (u, v) dynamics, these points are $(1, 0)$ and $(1/c, 0)$ and $(0, 1)$ and $(0, 1/b)$.) Does it matter biologically whether or not the lines intersect in the second or fourth quadrant when $c > 1$ and $b < 1$? Are there other cases?

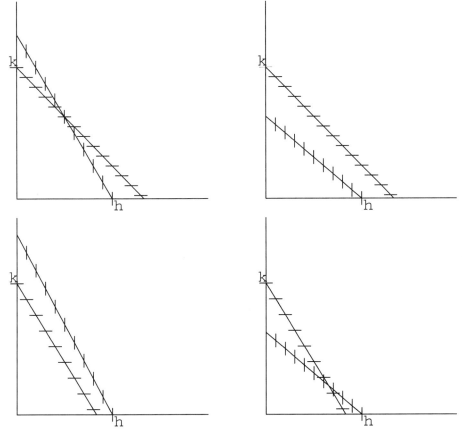

FIGURE 41.7: Four Kinds of Competition between Species

HINT 41.8. *Analytical Proofs*
*The computer program **LocalStability** performs the computations necessary to use the Local Stability Theorem for Continuous Dynamical Systems on these competition equations. This offers an analytical alternative to the simple geometric solution of the problem given in the previous exercise.*

41.2. Cooperation between Species

There is one small chain of idyllic islands on Sogapalag that are unique. Grass and bushes are present in perfect balance with rabbits and sparrows and they have no weasels. The rabbits fertilize the bushes in the process of eating the excess berries that the sparrows knock off the bushes. The fertilization makes the bushes more fertile and helps the sparrows, while the berries supplement the rabbits' diet. The dynamics are described by the equations:

$$\frac{dx}{dt} = mx\left(1 - \frac{x}{h} + f\frac{y}{k}\right)$$
$$\frac{dy}{dt} = ny\left(1 + g\frac{x}{h} - \frac{y}{k}\right)$$

where f and g are coefficients for cooperation. We may also write the equations in terms of the natural un-cooperative carrying capacities by a change of variables,

$$u = \frac{x}{h} \quad \text{and} \quad v = \frac{y}{k}$$
$$\frac{du}{dt} = \frac{1}{h}\frac{dx}{dt} \quad\quad \frac{dv}{dt} = \frac{1}{k}\frac{dy}{dt}$$

The cooperation equations become

$$\frac{1}{h}\frac{dx}{dt} = m\frac{1}{h}x\left(1 - \frac{x}{h} + f\frac{y}{k}\right) \quad \Leftrightarrow \quad \frac{du}{dt} = mu(1 - u + fv)$$
$$\frac{1}{k}\frac{dy}{dt} = n\frac{1}{k}y\left(1 + g\frac{x}{h} - \frac{y}{k}\right) \quad \Leftrightarrow \quad \frac{dv}{dt} = nv(1 + gu - v)$$

HINT 41.9. *Locate and characterize all nontrivial equilibrium points of the cooperation equations. Show that there is a positive equilibrium at (x_e, y_e) when $1 > fg$,*

$$\frac{x_e}{h} = \frac{1+f}{1-fg} \quad \text{and} \quad \frac{y_e}{k} = \frac{1+g}{1-fg}$$

Try some mathematical experiments with various values of the parameters. Try modest cooperation with $f = 1/4$ and $g = 1/2$ and extreme cooperation with $f = 1.5$ and $g = 1.25$

HINT 41.10. *Make compass heading charts like the ones in the text to show that there are two qualitatively different levels of cooperation between the species of this model. Modest cooperation raises both species above the natural uncooperative carrying capacity of both species, while larger amounts of mathematical cooperation lead to a biologically unrealistic case. What is the biological interpretation of both parameters satisfying $f > 1$ and $g > 1$?*

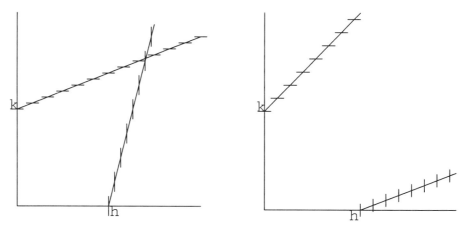

FIGURE 41.11: Two Kinds of Cooperation between Species

PROJECT 42

Sustained Harvest of Sei Whales

This project uses a discrete dynamical model proposed by J.R. Beddington in the *Report of the International Whaling Commission* (1978) to model the species of "Sei whales." We do not attempt to derive the model dynamics, but only investigate its consequences. In the case of our Sei whale model, we are primarily concerned with "whaling quotas."

42.1. Carrying Capacity, Environmental and Mathematical

Beddington's discrete dynamical model of Sei Whale growth is quite complicated. You need not try to justify it completely, but you should show that you understand what the main terms mean. We begin with an mathematical explanation of "carrying capacity" in terms of simpler models of population growth.

First, recall the discussion of algal growth in Section 3.3 of the main text. The simplest equation describing the growth of a population is the exponential growth model, where the continuous time rate of change of the number of individuals is given by

$$\frac{dN}{dt} = r\,N \quad \text{or, per capita,} \quad \frac{1}{N}\frac{dN}{dt} = r, \quad \text{a constant}$$

for r defined as the per capita growth constant of the population. This model simply says that each "mother" N has "babies" at the continuous rate r.

The equation above describes instantaneous growth. It gives a good model of the initial growth of algae that are all reproducing at different times. Sei whales, however, birth a calf only once every 9 years, resulting in a "time lag" in growth. Moreover, they do this in a discrete "calving season." To represent this situation accurately, more complicated mathematics must be developed. We need discrete time steps to correspond to the calving season.

The differential equation is approximately

$$\frac{N[t+\delta t] - N[t]}{\delta t} \approx r\,N[t] \quad \text{or} \quad N[t+\delta t] \approx N[t] + (r\cdot\delta t)\,N[t]$$

for time steps of size δt. In single unit time steps, this corresponds to the discrete dynamical system or discrete exponential growth law

$$p[t] = p[t-1] + \alpha\,p[t-1]$$

The growth in the population from year $t-1$ to year t is $p[t] - p[t-1]$, given by the term

$$\alpha \, p[t-1] \quad \text{or, per capita, by the constant} \quad \alpha$$

This simply says that each member of the population $p[t-1]$ at time $t-1$ gives birth to α new members at the start of year t. (Typically, α is a fraction, for example, $\alpha = 1/10$ means that each whale has one baby every 10 years, rather than a tenth of a baby each year.)

This model has the unrealistic feature that if $\alpha > 0$, the population explodes to infinity (whereas if $\alpha < 0$, the population cannot reproduce itself and becomes extinct. The symbolic solution, given in Chapter CD20 of the main text, is the exponential function $p[t] = p_0 (1 + \alpha)^t$.) In the long run, such a model cannot describe actual populations - it tends to infinity and so would lead to overpopulation when $\alpha > 0$. (Compare this with the results in the computer program **ExpGrth** from Chapter 28 of the main text, changing the 2 of the program to $(1 + \alpha)$.)

Overpopulation does not occur due to the limiting effects of food and space. Another term can be added to the population growth equation that depends on the carrying capacity of the environment. A simple "logistic term" gives us the discrete time equation

$$p[t] = p[t-1] + \alpha \, p[t-1] \left(1 - \frac{p[t-1]}{k}\right)$$

with the per capita growth

$$\alpha \left(1 - \frac{p[t-1]}{k}\right)$$

The parameter k is the mathematical counterpart of the environment's carrying capacity for the population. If $p[t-1]$ is small compared with k, then $p[t-1]/k \approx 0$ and the per capita growth is approximately $\alpha (1 - P[t-1]/k) \approx \alpha$. As $p[t-1]$ becomes closer to k, $p[t-1] \uparrow k$, the fraction $p[t-1]/k \uparrow 1$, so the expression for per capita growth $\alpha (1 - p[t-1]/k)$ tends toward zero, stopping the growth. If $p[t-1]$ becomes bigger than k, then $\alpha (1 - p[t-1]/k)$ is negative and the population declines.

Example CD20.3 and Exercise CD20.4.2 in the main text show that the simple logistic system stably approaches the carrying capacity when $0 < \alpha < 2$ and the initial population is not too far from k.

Beddington's model has the nonlinear term

$$0.0567 \left(1 - \left(\frac{p[t-9]}{k}\right)^{2.39}\right)$$

that is analogous to the logistic term. When $p[t-9] = k$, this term is zero.

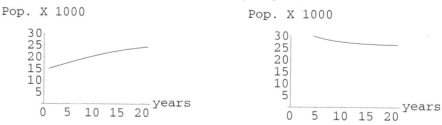

FIGURE 42.1: Changes in the Population

HINT 42.2. *Investigate the population model*

$$p[t] = p[t-1] + p[t-1]\left(0.0567\left(1 - (\frac{p[t-1]}{k})^{2.39}\right)\right)$$

*with $k = 26$. This model is similar to Beddington's model, but without the 8-year delay in births. Use **FirstDynSys** for experiments and apply the conditions of Theorem CD20.4 to test symbolically for local stability.*

42.2. The Actual Carrying Capacity

The actual environment's carrying capacity for Sei whales needs to be measured in the actual environment. That value becomes the value of k in our model. Our estimates of the past population of Sei whales is contained in the program **WhaleHelp**. A plot of that data is shown in Figure 42.3.

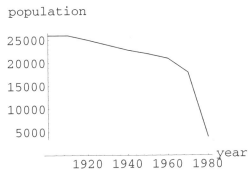

FIGURE 42.3: Estimated Sei Whale Population of the Indian Ocean

Beddington's model describes the whale population p at time t as a function of time. It does this by giving a discrete dynamical system for $p[t]$ in terms of $p[t-1]$ and $p[t-9]$. The new feature here, compared with Chapter CD20 of the main text, is that births depend on the population 9 years ago. It takes that long for a Sei whale to birth a calf.

new population = old population − annual deaths + annual births
− hunted who wouldn't have died from natural causes

$$p[t] = p[t-1] - 0.06p[t-1] + p[t-9]\left(0.06 + 0.0567\left(1 - (\frac{p[t-9]}{k})^{2.39}\right)\right)$$
$$- 0.94h$$
$$p[t] = 0.94\,p[t-1] + \text{Births}(p[t-9]) - 0.94\,h$$

The parameter h in the last term represents the number of whales that are hunted in the year. If $h = 1000$ whales were hunted, 6% of these would have died of natural causes anyway, so we deduct only $.94\,h$. The parameter k is the environmental carrying capacity.

HINT 42.4. *Begin your project with a biological explanation of the above equation. What is the formula for the net growth in the population at the beginning of year t? What is the net per capita growth?*

*Estimate the carrying capacity from Figure 42.3 (or the data in **WhaleHelp**). You will use your estimate as k in the model.*

Here is a test case to use in helping us set whaling quotas from the model. Suppose there is no hunting and the population has held at a constant level for 9 years, $p = p[t-1] = p[t-9]$, what is the total change in the annual population for next year? Use the computer to plot the (un-hunted) annual change as a function of $p = p[t-1] = p[t-9]$ for values of p from 0 to $1.2\,k$.

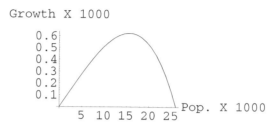

FIGURE 42.5: Natural Change versus P

Now, consider the case in which the harvest levels are set $h \neq 0$. Suppose the whale stock reaches some hunted equilibrium value, $p_H < k$. If the population is at this constant population, $p_H = p[t] = p[t-1] = p[t-9]$, it will remain at this level. When this equilibrium exists, the corresponding harvest quota, h, is called a sustainable yield. The maximum sustainable yield is then defined as the largest yield for which the function has a nonzero equilibrium value.

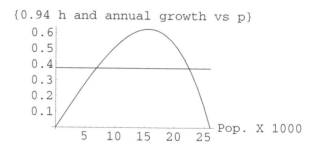

FIGURE 42.6: Change and .94 Hunting

HINT 42.7. *Explain the meaning of the term "sustainable yield." Use the computer to plot both the natural change and the level $0.94\,h$ on the same plot. For example, plot the case where $h = 400$ whales are hunted per year. When is the total change due to both hunting and natural causes positive? Where does the difference (net natural change $-0.94\,h$) appear on the combined graph?*

It is important to find the maximum sustainable yield. When hunting is set at that level, whalers get the most whales possible while maintaining a population that can be hunted in the future.

HINT 42.8. *What is the maximum sustainable yield? Is the annual growth ever positive when $h = 700$? Is it ever positive if hunting quotas are set exactly at the MSY? What is the highest h so that total change with hunting has an equilibrium?*

Biologists and mathematicians have found that modeling the population of marine organisms like whales is a very complicated, uncertain process. The parameters cannot be measured accurately. It is therefore hard to estimate the biologically maximum sustainable yield, as well as to test the validity of a given model. For this reason, it has been suggested that the harvest quota should be set at a fraction of the mathematically maximum sustainable yield derived from a model such as this one.

HINT 42.9. *With hunting set at the maximum sustainable yield, determine what will happen after several years of hunting at this level. What would happen if the MSY had been miscalculated, and was actually lower than the level set for hunting?*

HINT 42.10. *Consider a situation in which hunting was set at 0.85 of the maximum sustainable yield - how many equilibrium values exist for this hunting? What are the characteristics of these equilibrium values? How does the population behave if hunting is maintained at this level for many years?*

Simply looking a decade or two down the road isn't sufficient to ensure that the population will survive over centuries. The future of the Sei whale population is in your hands - you must determine how the population will be affected by various hunting levels, which should be described in relation to the maximum sustainable yield.

The following exercise is a suggestion for investigations that will be helpful in understanding the dynamics of the model. In addition, you are encouraged to develop other explorations that will better characterize the Sei whale population's dynamics, and so provide more information on how the population may be saved. Good luck, and happy hunting! May your great-grandchildren hunt the grandchildren of our friend the Sei whale and each sustain the other.

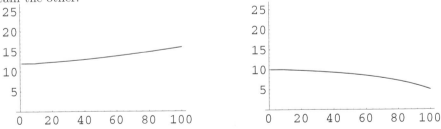

FIGURE 42.11: 100 Years of Hunting with Different Starting Populations

HINT 42.12. *Look into the future mathematically – what will the Sei whale population be like 400 years from now? How do different hunting levels and or different initial populations affect this future population? Under what conditions will the population become extinct? Remember that if some poaching takes place, the initial population could be suppressed below the hunted equilibrium. There is a computer program **WhaleHelp** to help you with your computations. Will the whale recover from its 1980 level?*

Chapter 14

Derivations with Vectors

Vector geometry and the meaning of the derivative can be combined to describe the motion of many real objects. This chapter offers a few examples of this powerful combination.

PROJECT 43

Wheels Rolling on Wheels

A small wheel rolls around a large fixed wheel without slipping. What path does a point on the edge of the small wheel trace out as it goes? You can answer this quaestion with vectors and an understanding of radian measure. The Spirograph toy makes such plots, called "cycloids." Figure 43.1 is a sample path.

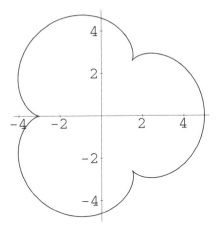

FIGURE 43.1: An epicycloid

This is not trivial. Vectors help, and here is a special case to show you why. Suppose the large wheel has radius 3 and the small wheel has radius 1. How many turns must the small wheel make while traveling once around the large wheel? This is a famous question because it appeared on a national college entrance test as a multiple-choice question. The reason it is famous is that the correct answer was not among the choices.

Here is the obvious solution: The big wheel has three times the circumference of the small wheel (you can use $C = 2\pi r$), so the small one must mark off three revolutions to cover the same distance and travel all the way around the large one. Obvious, but wrong. The small wheel turns four times. Run the computer animation in **EpiCyAnimate** to see for yourself (Figure 43.2).

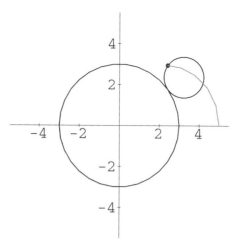

FIGURE 43.2: **EpiCyAnimate**

43.1. Epicycloids

We will use vectors to analyze the motion of a point on the little wheel. This will give us parametric equations for the whole path but also show us why the small wheel turns four times in the 3-to-1 ratio case. You will need to know that the length of a circular segment of angle θ and radius r is

$$\text{arc length} = r\,\theta$$

For example, the length all the way around a circle of radius r is what we get when we go an angle $\theta = 2\pi$, so $C = 2\pi\,r$.

We begin with one vector $\mathbf{C}[\theta]$ pointing to the center of the small wheel. We know from the sine-cosine discussion that

$$\mathbf{C}[\theta] = \begin{bmatrix} c_1 \\ c_2 \end{bmatrix} = \begin{bmatrix} R\,\text{Cos}[\theta] \\ R\,\text{Sin}[\theta] \end{bmatrix}$$

if θ measures the angle from the x-axis and $R = r_b + r_s$, the sum of the radii of the large and small wheels.

We also have a second vector, $\mathbf{W}[\omega]$, pointing from the center of the small wheel to the tracing point on its rim. Again using sine-cosine, we have

$$\mathbf{W}[\omega] = \begin{bmatrix} w_1 \\ w_2 \end{bmatrix} = \begin{bmatrix} r_s\,\text{Cos}[\omega] \\ r_s\,\text{Sin}[\omega] \end{bmatrix}$$

where ω is the angle measured around the small wheel starting from the horizontal and r_s is the radius of the small wheel.

The vector $\mathbf{W}[\omega]$ is a displacement; it does not start at the origin. The position vector interpretation is simply that

$$\mathbf{X} = \mathbf{C} + \mathbf{W}$$

is the arrow from the origin to the reference point on the small wheel. This may be written in components as

$$x_1 = c_1 + w_1$$
$$x_2 = c_2 + w_2$$

where the components of **C** and **W** are given by the formulas above.

The two vectors start in the position shown in figure 43.3 with both pointing to the right of their centers.

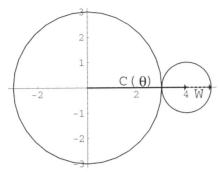

FIGURE 43.3: Starting Position

After a small increase in θ, the vectors look as in Figure 43.4.

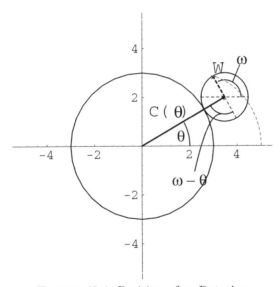

FIGURE 43.4: Position after Rotation

The length of portion along the big wheel where contact has been made is $r_b \theta$. The portion contacted along the small wheel is $r_s(\omega - \theta)$ because the contact point lags behind the original horizontal reference line. "No slipping" means that equal lengths are marked off along both wheels, so

$$r_b \theta = r_s (\omega - \theta)$$
$$r_s \omega = (r_b + r_s) \theta$$
$$\omega = \frac{r_b + r_s}{r_s} \theta$$

Consider the case where $r_b = 3$ and $r_s = 1$. When $\theta = 2\pi$ so that the small wheel has gone all the way around the large wheel, we have $\omega = \frac{3+1}{1} 2\pi = 4 \cdot 2\pi$. The small wheel turns four times a full turn.

The computer program **Cycloid** on our website contains the formulas for the epicycloid we have just derived and uses them to plot the curve. We want you to extend the idea of using parametric circles and vector addition together in the next exercise.

43.2. Cycloids

HINT 43.5. *The Cycloid*
A wheel of radius $r_s = 1$ rolls to the right along a straight line at speed v without slipping. The vector pointing to the center of the wheel has the form

$$\mathbf{C}[t] = \begin{bmatrix} c_1 \\ c_2 \end{bmatrix} = \begin{bmatrix} v \cdot t \\ ? \end{bmatrix}$$

where t is the time.

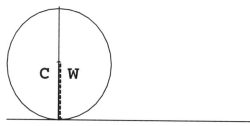

FIGURE 43.6: *Start of a Cycloid*

The vector pointing from the center of the wheel to a reference point on the wheel has the form

$$\mathbf{W}[\omega] = \begin{bmatrix} w_1 \\ w_2 \end{bmatrix} = \begin{bmatrix} -r \, \text{Sin}[\omega] \\ -r \, \text{Cos}[\omega] \end{bmatrix}$$

In what position does the wheel begin?

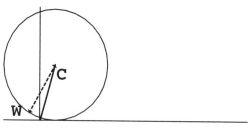

FIGURE 43.7: *After a Small Change*

The distance moved along the line at time t is $v \cdot t$. How much is the distance marked along the wheel in terms of ω? Equate these, assuming no slipping, and find ω in terms of t.

*Modify the computer program **Cycloid** to plot this curve, known as a cycloid.*

HINT 43.8. *Make an integral computation of the area under one loop of the cycloid.*

43.3. Hypocycloids

EXPLORATION 43.9. *The Hypocycloid*
A small wheel of radius r_s rolls around the inside of a large fixed wheel of radius r_b. Use vectors to find parametric equations for the path of a point on the tip of the small wheel and plot the path using the computer. Three main things change compared with the epicycloid above: First, the small wheel rolls in the opposite direction to its center. You could account for this by taking

$$\mathbf{W}[\omega] = r_s \cdot \begin{bmatrix} \text{Cos}[-\omega] \\ \text{Sin}[-\omega] \end{bmatrix}$$

Why does a negative angle correspond to clockwise rotation? Second, the distance from the center of the big wheel to the center of the small wheel is no longer the sum of the radii. Third, the equation that says the same lengths are contacted along the large and small wheels (or no slipping occurs) is different. This time the opposite rotation causes angles to add in the length equation along the small wheel.

If the large wheel has radius 3 and the small wheel has radius 1, how many turns does the small wheel make when going around the large wheel once for the epicycloid? For the hypocycloid? (What is ω when $\theta = 2\pi$?)

(Optional) What happens if you use the epicycloid program and equations with $r = -1$ or a negative r_s in general? Try this with computer, and then show that the curve really is a hypocycloid but corresponds to drawing the curve with a Spirograph starting the small wheel in a different position.

PROJECT 44

The Perfecto Skier

This project investigates the 3-D effect of sliding down a slippery slope. The component of the force due to graivty acts opposite the gradient, but motion under this acceleration does not produce gradient trajectories. You don't just ooze to the bottom, but zoom right on by. Problem 18.2 of the core text found the gradient trajectories. Exercise 17.7.6 found the tangential component of a force acting on a plane. We begin with a quick reminder about those problems.

Perfecto Ski Bowl has the shape of the surface

$$z = x^2 + 2y^2$$

HINT 44.1. *Show that the normal vector to the surface of Perfecto above the point (x, y) is*

$$\begin{bmatrix} -2x \\ -4y \\ 1 \end{bmatrix}$$

A skier falls somewhere on the slope above (x_0, y_0) with initial velocity zero but is without control of direction; his edges are in the air.

The skier's weight vector is

$$\begin{bmatrix} 0 \\ 0 \\ -mg \end{bmatrix}$$

where g is Newton's gravitational constant ($g = 9.8$ in mks units). The mountain counteracts the force along its normal, but the rest of his weight vector acts down the mountain.

HINT 44.2. *Show that the force pulling the skier down the mountain when he is at the point above (x, y) is*

$$\mathbf{F}_T = \frac{-4mg}{4x^2 + 16y^2 + 1} \begin{bmatrix} x/2 \\ y \\ x^2 + 4y^2 \end{bmatrix}$$

Newton's "F = m a" law says that the mass times the acceleration of the skier, $\mathbf{a} = \frac{d^2\mathbf{X}}{dt^2}$, equals the total force on the skier,

$$m\mathbf{a} = \mathbf{F}_T + \mathbf{F}_M$$

$$m\frac{d^2\mathbf{X}}{dt^2} = \mathbf{F}_T + \mathbf{F}_M$$

$$m\begin{bmatrix} \frac{d^2x}{dt^2} \\ \frac{d^2y}{dt^2} \end{bmatrix} = \frac{-4\,m\,g}{4\,x^2 + 16\,y^2 + 1}\begin{bmatrix} x/2 \\ y \\ x^2 + 4\,y^2 \end{bmatrix} + \mathbf{F}_M$$

where $\mathbf{F}_M$ is the force supplied by the mountain. If we assume that the slope is perfectly slippery, this force will be perpendicular to the mountain and act just to keep the skier on the curves of the mountain.

44.1. The Mountain's Contribution

A skier at rest will only have the tangential component of gravity acting on him, but in motion, the mountain must contribute a force. We want you to see a simple case of this force without worrying about gravity.

The vector function

$$\mathbf{X}[t] = \begin{bmatrix} r\,\mathrm{Cos}[\omega\,t] \\ r\,\mathrm{Sin}[\omega\,t] \\ 0 \end{bmatrix}$$

moves at constant speed around the circle of radius r, given implicitly by $x^2 + y^2 = r^2$ and $z = 0$.

HINT 44.3. *Compute $|\mathbf{X}|$, $\left|\frac{d\mathbf{X}}{dt}\right|$ and $\left|\frac{d^2\mathbf{X}}{dt^2}\right|$ and show that they are all different constants.*

Suppose that we want the speed of the motion to be s. Show that we need to take $\omega = s/r$, so

$$\frac{d^2\mathbf{X}}{dt^2} = -\left(\frac{s}{r}\right)^2 \mathbf{X}$$

The acceleration of a circular "mountain" on a "skier" at constant speed s is s^2/r in the direction $-\frac{1}{r}\mathbf{X}$.

It is easiest to represent our mountain implicitly

$$h[\mathbf{X}] = j, \text{ a constant}$$

For example, for Perfecto, we can take the function $h[x,y,z] = x^2 + 2y^2 - z$ and the implicit equation, $h[x,y,z] = 0$. The reason we want to use this form is so we have a simple way to find a normal to the mountain. In this case, the gradient

$$\nabla h[x,y,z] = \begin{bmatrix} \frac{\partial h}{\partial x} \\ \frac{\partial h}{\partial y} \\ \frac{\partial h}{\partial z} \end{bmatrix} \text{ is a vector perpendicular to } h[x,y,z] = j$$

HINT 44.4. *Verify that ∇h is perpendicular to Perfecto Ski Bowl when $h = x^2 + 2y^2 - z$.*

HINT 44.5. *If **F** is any vector, derive a formula for the component of **F** that is tangent to the mountain* $h[\mathbf{X}] = j$,

$$\mathbf{F}_T = \mathbf{F} - \frac{\mathbf{F} \bullet \nabla h}{\nabla h \bullet \nabla h} \nabla h$$

Test your formula in the case $\mathbf{F} = (0, 0, -mg)$ *and* $h[x, y, z] = x^2 + 2y^2 - z$.

The force due to the mountain will be a scalar multiple of the perpendicular vector, $\mathbf{F}_M = \lambda \nabla h$, and

$$m \frac{d^2 \mathbf{X}}{dt^2} = \mathbf{F}_T + \lambda \nabla h$$

We want to compute λ, knowing that it somehow depends on the curvature of the mountain and the skier's speed.

HINT 44.6. *Let* $\mathbf{X}[t]$ *be any parametric function and suppose it lies on the mountain,*

$$h[\mathbf{X}[t]] = j$$

Differentiate this equation once with respect to t and show that

$$\nabla h \bullet \frac{d\mathbf{X}}{dt} = 0$$

What does this equation say about the direction of the velocity of the motion $\mathbf{X}[t]$?

The equation above can be written in components as

$$\frac{\partial h}{\partial x} \frac{dx}{dt} + \frac{\partial h}{\partial y} \frac{dy}{dt} + \frac{\partial h}{\partial z} \frac{dz}{dt} = 0$$

If we differentiate this equation with respect to t again using the Product Rule and Chain Rule, we obtain

$$\frac{\partial^2 h}{\partial x^2} \left(\frac{dx}{dt}\right)^2 + 2 \frac{\partial^2 h}{\partial x \partial y} \left(\frac{dx}{dt}\right) \left(\frac{dy}{dt}\right)$$
$$+ \frac{\partial^2 h}{\partial y^2} \left(\frac{dy}{dt}\right)^2 + 2 \frac{\partial^2 h}{\partial y \partial z} \left(\frac{dy}{dt}\right) \left(\frac{dz}{dt}\right)$$
$$+ \frac{\partial^2 h}{\partial z^2} \left(\frac{dz}{dt}\right)^2 + 2 \frac{\partial^2 h}{\partial z \partial x} \left(\frac{dz}{dt}\right) \left(\frac{dx}{dt}\right)$$
$$+ \frac{\partial h}{\partial x} \frac{d^2 x}{dt^2} + \frac{\partial h}{\partial y} \frac{d^2 y}{dt^2} + \frac{\partial h}{\partial z} \frac{d^2 z}{dt^2} = 0$$

Notice that the last line of this derivative contains the dot product

$$\nabla h \bullet \frac{d^2 \mathbf{X}}{dt^2}$$

We summarize the computation in matrix notation

$$\nabla h \bullet \frac{d^2 \mathbf{X}}{dt^2} = -\frac{d\mathbf{X}}{dt}^t \cdot \begin{bmatrix} \frac{\partial^2 h}{\partial x^2} & \frac{\partial^2 h}{\partial x \partial y} & \frac{\partial^2 h}{\partial z \partial x} \\ \frac{\partial^2 h}{\partial x \partial y} & \frac{\partial^2 h}{\partial y^2} & \frac{\partial^2 h}{\partial y \partial z} \\ \frac{\partial^2 h}{\partial x \partial z} & \frac{\partial^2 h}{\partial z \partial y} & \frac{\partial^2 h}{\partial z^2} \end{bmatrix} \cdot \frac{d\mathbf{X}}{dt}$$

$$\nabla h \bullet \frac{d^2 \mathbf{X}}{dt^2} = -\frac{d\mathbf{X}}{dt}^t \cdot h^{(2)}[\mathbf{X}] \cdot \frac{d\mathbf{X}}{dt}$$

where $h^{(2)}[\mathbf{X}]$ is the matrix of second partial derivatives of the constraint function $h[\mathbf{X}]$. Notice that this term involves the velocity and second partials related to the curves on the mountain.

44.2. Gravity and the Mountain

We use this fact about any motion on $h[X[t]] = j$ together with the fact that the tangential force is perpendicular to ∇h to derive a formula for the scalar λ:

dot ∇h on Newton's law

$$\nabla h \bullet m \frac{d^2 \mathbf{X}}{dt^2} = \nabla h \bullet \mathbf{F}_T + \nabla h \bullet \lambda \nabla h$$

use perpendicularity

$$m \nabla h \bullet \frac{d^2 \mathbf{X}}{dt^2} = \lambda |\nabla h|^2$$

and, from above,

$$\nabla h \bullet \frac{d^2 \mathbf{X}}{dt^2} = -\frac{d\mathbf{X}}{dt}^t \cdot h^{(2)}[\mathbf{X}] \cdot \frac{d\mathbf{X}}{dt}$$

HINT 44.7. *Solve these two equations for λ:*

$$\lambda = -\frac{m}{|\nabla h[\mathbf{X}]|^2} \frac{d\mathbf{X}}{dt}^t \cdot h^{(2)}[\mathbf{X}] \cdot \frac{d\mathbf{X}}{dt}$$

Test your formula on the case of constant speed s motion around the circle $x^2 + y^2 = r^2$, $z = 0$. Newton's law with no tangential force becomes

$$m \frac{d^2 \mathbf{X}}{dt^2} = \lambda \nabla h[\mathbf{X}]$$

Now compute the specific quantities:

$$h[x, y, z] = x^2 + y^2, \quad \nabla h = ??, \quad |\nabla h|^2 = 4 \ ?^2$$

$$h^{(2)} = \begin{bmatrix} 2 & 0 & 0 \\ 0 & 2 & 0 \\ 0 & 0 & 0 \end{bmatrix}, \quad \frac{d\mathbf{X}}{dt}^t \cdot h^{(2)}[\mathbf{X}] \cdot \frac{d\mathbf{X}}{dt} = 2 \ ?^2$$

HINT 44.8. *Recall that the force on the skier is $\frac{m s^2}{r}$ in the direction $-\frac{1}{r}\mathbf{X}$. Simplify the expressions in this special case and show that the formula $m \frac{d^2\mathbf{X}}{dt^2} = \lambda \nabla h[\mathbf{X}]$ agrees with this, $\lambda \nabla h[\mathbf{X}] = -m \frac{s^2}{r^2}\mathbf{X}$.*

44.3. The Pendulum as Constrained Motion

We may view the pendulum (described in Project 46) as a motion on a circle
$$x^2 + y^2 = L^2$$
so $h[x, y, z] = x^2 + y^2$. The only applied force is still gravity which we take as $F = (0, -mg, 0)$.

HINT 44.9. *Show that the differential equation describing the motion of a pendulum of length L with gravitational constant g may be written:*
$$\frac{dx}{dt} = u$$
$$\frac{dy}{dt} = v$$
$$\frac{du}{dt} = \frac{g}{L^2} x y - \frac{u^2 + v^2}{L^2} x$$
$$\frac{dv}{dt} = -\frac{g}{L^2} x^2 - \frac{u^2 + v^2}{L^2} y$$

44.4. The Explicit Surface Case

Suppose that the surface constraining the motion can be expressed as an explicit function of x and y,
$$z = k[x, y]$$
The constraint $h[x, y, z] = j$ can be written $h[x, y, z] = k[x, y] - z = 0$ for the general formulas.

HINT 44.10. *Show that*
$$\nabla h = \begin{bmatrix} \frac{\partial k}{\partial x} \\ \frac{\partial k}{\partial y} \\ -1 \end{bmatrix} \quad \text{and} \quad |\nabla h|^2 = \left(\frac{\partial k}{\partial x}\right)^2 + \left(\frac{\partial k}{\partial y}\right)^2 + 1$$

and compute this quantity for the Perfecto Ski Bowl function.

HINT 44.11. *Show that*
$$h^{(2)} = \begin{bmatrix} \frac{\partial^2 k}{\partial x^2} & \frac{\partial^2 k}{\partial x \partial y} & 0 \\ \frac{\partial^2 k}{\partial y \partial x} & \frac{\partial^2 k}{\partial y^2} & 0 \\ 0 & 0 & 0 \end{bmatrix}$$

and

$$[u \ v \ w] \cdot h^{(2)}[\mathbf{X}] \cdot \begin{bmatrix} u \\ v \\ w \end{bmatrix} = [u \ v] \begin{bmatrix} \frac{\partial^2 k}{\partial x^2} & \frac{\partial^2 k}{\partial x \partial y} \\ \frac{\partial^2 k}{\partial y \partial x} & \frac{\partial^2 k}{\partial y^2} \end{bmatrix} \begin{bmatrix} u \\ v \end{bmatrix} = u^2 \frac{\partial^2 k}{\partial x^2} + 2 u v \frac{\partial^2 k}{\partial x \partial y} + v^2 \frac{\partial^2 k}{\partial y^2}$$

and compute this quantity for the Perfecto Ski Bowl function.

HINT 44.12. *Show that when* $\mathbf{F} = (0, 0, -mg)$,

$$\mathbf{F}_T = mg \begin{bmatrix} 0 \\ 0 \\ -1 \end{bmatrix} - \frac{mg}{\left(\frac{\partial k}{\partial x}\right)^2 + \left(\frac{\partial k}{\partial y}\right)^2 + 1} \begin{bmatrix} \frac{\partial k}{\partial x} \\ \frac{\partial k}{\partial y} \\ -1 \end{bmatrix}$$

and compute this quantity for the Perfecto Ski Bowl function.

HINT 44.13. *Let* $P = \left(\frac{\partial k}{\partial x}\right)^2 + \left(\frac{\partial k}{\partial y}\right)^2 + 1$, $\frac{dx}{dt} = u$, $\frac{dy}{dt} = v$, $\frac{dz}{dt} = w$, *and* $Q = u^2 \frac{\partial^2 k}{\partial x^2} + 2uv \frac{\partial^2 k}{\partial x \partial y} + v^2 \frac{\partial^2 k}{\partial y^2}$. *Show that the equations of motion on* $z = k[x, y]$ *may be written*

$$\frac{du}{dt} = -\left(\frac{g}{P} + \frac{Q}{P}\right) \frac{\partial k}{\partial x}$$

$$\frac{dv}{dt} = -\left(\frac{g}{P} + \frac{Q}{P}\right) \frac{\partial k}{\partial y}$$

$$\frac{dw}{dt} = \frac{g}{P} + \frac{Q}{P} - g$$

and write these equations for the Perfecto Ski Bowl function.

Notice that the z variable does not appear in the equations for x, y, u, and v. This means that we could solve the pair of second-order equations for x and y, or $\frac{dx}{dt} = u$, $\frac{dy}{dt} = v$, and

$$\frac{du}{dt} = -\left(\frac{g}{P} + \frac{Q}{P}\right) \frac{\partial k}{\partial x}$$

$$\frac{dv}{dt} = -\left(\frac{g}{P} + \frac{Q}{P}\right) \frac{\partial k}{\partial y}$$

together with initial conditions and find functions $x[t]$ and $y[t]$. We could then simply let $z[t] = h[x[t], y[t]]$.

Now put it all together:

EXPLORATION 44.14. *Use the phase variable trick from Chapter 23 of the core text to write the equations of motion for the skier as a differential equation:*

$$\frac{dx}{dt} = u$$

$$\frac{du}{dt} = \ldots$$

$$\vdots$$

In a program, you can "fill in the blanks" in simple steps using computer derivatives and the formulas above. Solve your equations and plot the path of the skier.

Use the **PerfectoHelp** program from our website to solve these equations for several choices of the initial conditions. Does the skier stop at the bottom?

PROJECT 45

Low-Level Bombing

This project models low-level bombing with differential equations. This involves determining where a bomb lands when you drop it from a specific height and velocity, or when and where to release a bomb when you are given a specific, stationary target. The two ideas are Galileo's law of gravity and air friction due to the high speed of the bomb. They are combined with Newton's law, $\mathbf{F} = m\mathbf{A}$. Vectors play a role, because the air friction is an increasing function of speed, not velocity.

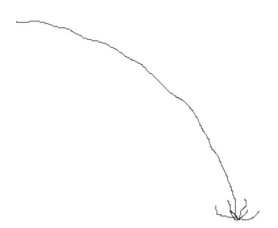

FIGURE 45.1: A Hypothetical Bomb Trajectory

BASIC ASSUMPTIONS

We place a coordinate system with its origin at the target with the horizontal x-axis aligned so the plane has positive x-vleocity. We will use mks units, so the gravitational constant $g = 9.8$.

Essential Variables

> t Time, measured in seconds
> $x(t)$ Horizontal bomb position, distance measured along the ground in meters
> $y(t)$ Vertical bomb position, measured vertically in meters

Auxiliary Variables

These variables make the translation of the information easier but are defined in terms of the main variables. Two of them are "phase variables."

$$u = \frac{dx}{dt} \qquad \text{Horizontal velocity } (m/sec)$$

$$v = \frac{dy}{dt} \qquad \text{Vertical velocity } (m/sec)$$

$$s = |\mathbf{V}| \qquad \text{Total speed of the bomb } (m/sec)$$

$$= \sqrt{u^2 + v^2} = \text{Sqrt}[u^2 + v^2]$$

Note that $\mathbf{X}(t) = \begin{bmatrix} x(t) \\ y(t) \end{bmatrix}$ is the position vector of the bomb, $\mathbf{V} = \begin{bmatrix} u(t) \\ v(t) \end{bmatrix} = \begin{bmatrix} \frac{dx}{dt} \\ \frac{dy}{dt} \end{bmatrix}$ is the velocity vector, and its length is the speed of the bomb. Acceleration is the vector $\mathbf{A}(t) = \begin{bmatrix} \frac{d^2x}{dt^2} \\ \frac{d^2y}{dt^2} \end{bmatrix} = \frac{d\mathbf{V}}{dt}$.

We will model air friction by assuming that the force due to air friction is proportional to a power of the speed and acts in the direction opposite the velocity vector, $|\mathbf{F}_{air}| = ks^p$, with the same direction as $-\mathbf{V}$.

Parameters

> m Mass of the bomb (kg)
> k Air resistance constant, coefficient of friction due to air resistance
> p Power of the speed in air resistance force

Remember that parameters are simply "variables that are held constant." In this case, they are the particular physically measurable quantities that characterize a particular bomb. They are "constant" for each bomb but vary from bomb to bomb.

Initial Conditions

> $\begin{bmatrix} x_0 \\ y_0 \end{bmatrix}$ - Initial position, position vector from which the bomb was originally dropped
>
> $\begin{bmatrix} u_0 \\ v_0 \end{bmatrix}$ - Initial velocity, velocity vector when the plane releases the bomb

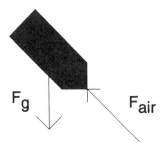

FIGURE 45.2: A Vector Force for Gravity and a Vector Force for Air Resistance

AIR RESISTANCE

One of the two forces acting on the bomb as it falls is air resistance, $\mathbf{F}_{air}$. This force acts to slow the bomb opposite to the direction in which the bomb tries to travel. It acts in an opposite direction to the velocity vector at any given time, as Figure 45.2 illustrates.

HINT 45.3. *The Air's Force*
1) What is the unit vector in the direction of $\mathbf{V}$, the velocity vector? Express your answer in terms of u, v, and s.
2) What is the unit vector in the direction opposite $\mathbf{V}$, or in the direction of $\mathbf{F}_{air}$?
3) Using the results from the two previous exercises, show that

$$\mathbf{F}_{air} = -ks^{p-1}\begin{bmatrix} u \\ v \end{bmatrix}$$

Verify that $|\mathbf{F}_{air}| = ks^p$ and that the direction of $\mathbf{F}_{air}$ is opposite the direction of $\mathbf{V}$.

The other force acting on the bomb is due to the weight of the bomb due to gravity. The force due to gravity varies according to mass and has magnitude mg, where g is the acceleration due to gravity, approximately equal to 9.8 meters/second^2.

HINT 45.4. *The Weight Vector*
What is the weight vector of the bomb as a vector in terms of the mass m and the acceleration due to gravity, g?

HINT 45.5. *The Total Force*
Now that we have the two forces acting on the bomb in vector form, compute the total force, $\mathbf{F}_{total}$, as a vector sum, $\mathbf{F}_{total} = \mathbf{F}_{air} + \mathbf{W}$. Express your answer in terms of u, v, m and g,

$$\mathbf{F}_{total} = \begin{bmatrix} ? \\ ?? \end{bmatrix}$$

HINT 45.6. $\mathbf{F} = m\mathbf{A}$
You should be familiar with Newton's law $\mathbf{F} = m\mathbf{A}$ from Chapter 10 of the text or from your physics course. This can be applied to our model, where $\mathbf{F}$ is the force vector, $\mathbf{F}_{total}$ the total force acting on the bomb, and $\mathbf{A}$ is the acceleration vector,

$$\mathbf{A} = \frac{d\mathbf{V}}{dt} = \begin{bmatrix} \frac{du}{dt} \\ \frac{dv}{dt} \end{bmatrix} = \begin{bmatrix} \frac{d^2x}{dt^2} \\ \frac{d^2y}{dt^2} \end{bmatrix}$$

Now rewrite this equation in terms of u, v, m and g solving for $\mathbf{A}$, $\mathbf{A} = \frac{1}{m}\mathbf{F}_{total}$,

$$\frac{d^2x}{dt^2} = \frac{du}{dt} = ?$$

$$\frac{d^2y}{dt^2} = \frac{dv}{dt} = ??$$

In order to use the **AccDEsoln** program to solve the equations, you must use the phase variable trick. This transforms the two second-order differential equations that you just derived into a system of four first-order differential equations.

HINT 45.7. *A First Order 4-D System*
Write $\mathbf{A} = \frac{1}{m}\mathbf{F}_{total}$ in terms of first derivatives and the phase variables.

$$\frac{dx}{dt} = u$$

$$\frac{dy}{dt} = v$$

$$\frac{du}{dt} = ?$$

$$\frac{dv}{dt} = ??$$

These four equations can be plugged into the **AccDEsol** program; however you also need the initial conditions. A plane flying horizontally at height h, a horizontal distance j before the target at speed s_i gives:

$$\begin{aligned}
x(0) &= -j &&\text{horizontal position}\\
y(0) &= h &&\text{vertical position}\\
u(0) &= s_i &&\text{horizontal speed of the bomber}\\
v(0) &= 0 &&\text{vertical speed of the bomber}
\end{aligned}$$

You will also need a specific value of the nonlinear power p in the air resistance term. In the Bungee project we used $p = 7/5$. Linear air resistance would be $p = 1$. Another simple nonlinear choice is $p = 2$. We would like you to try several choices of p. Each choice will require an associated constant of proportionality, k. Fortunately, we know the terminal velocity of the bombs from data at the testing center. When we drop these kinds of bombs vertically from a test tower, they speed up to approximately 89.4 meters per second and then continue to fall at that speed. Apparently, the air friction balances gravity at that speed.

HINT 45.8. *Measuring k from Terminal Velocity*
You can use simple algebra to compute the value of k. Since $\frac{du}{dt} = 0$ in a vertical drop, we have the equations

$$\frac{dy}{dt} = v$$

$$\frac{dv}{dt} = -g - \frac{k}{m}\,\text{sign}(v)|v|^p$$

What does "terminal velocity" say about $\frac{dv}{dt}$? Solve this equation for k in terms of p and compute the specific values for $p = 1, 7/5, 2$.

45.1. Significance of Vector Air Resistance

The vector nature of the air resistance means that a bomb released at high speed will have a big vector force opposite to its direction of motion. If the bomb is inclined downward, the resistance will therefore have an upward component. The bomb won't fall if this component of resistance equals its weight. Of course, the resistance slows the bomb down horizontally, so it does not fly along with the bomber for long. Once it slows down, weight exceeds the upward component of resistance and it falls faster and faster until the vertical resistance again builds up.

EXPLORATION 45.9. *Low-Flying Bombs*
Run some computer experiments with bombers flying 350 mph (be careful with units), dropping bombs from various initial heights. Use several values of the resistance power p and its associated k. Show that the flying effect of bombs is most important in hitting a target if you wish to bomb from a low level.

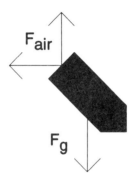

FIGURE 45.10: Force Components

Now that you have a good understanding of how this model works and what the equations mean both graphically and mathematically, it is time to put the computer to work.

EXPLORATION 45.11. *Hitting the Target*
You are flying at altitude h at velocity s_i and decide to drop a bomb. Where do you release the bomb so that it hits the target at (0,0)? Use several values of p. How much room for error do you have in releasing the bomb so that you will still hit within a meter of your target? How does varying your altitude and initial velocity affect the room for error that you have?

It is also interesting to follow the path of the bomb. This can be graphed with the computer's help. In addition, you can follow the descent by following the angle the tangent line to the bomb's path forms with the ground. This can tell you the angle at which the bomb hits, or if the bomb nearly "floats" when first released.

EXPLORATION 45.12. *Determine the angle at impact of the bomb. How do changes in h and s_i affect this?*

PROJECT 46

The Pendulum

The simple pendulum has been the object of much mathematical study. A device as simple as the pendulum may not seem to warrant much mathematical attention, but the study of its motion was begun by some of the greatest scientists of the 17th century. The reason was that the pendulum allows one to keep time with great accuracy. The development of pendulum clocks made sea travel safer during a time when new worlds in the Americas were just being developed and discovered. The scientist Christian Huygens studied the pendulum extensively from a geometric standpoint, and Issac Newton's theory of gravitation made it possible to study the pendulum from an analytic standpoint, which is what we will do in this project. The compound pendulum is still studied today because of its complicated "chaotic" behavior.

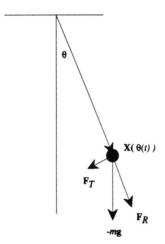

FIGURE 46.1: The Pendulum

A simple pendulum is a small mass suspended from a light rigid rod. Ideally the mass will be small enough so that it will not deform the rod, but the rod should be light so that we can neglect its mass. (We will consider what happens if the rod stretches later in the project.) If we displace the bob slightly and let it go, the pendulum will start swinging. We will assume at first that there is no friction. In this idealized situation, once the pendulum is started, it will never stop. The only force acting on the pendulum is the force of gravity.

46.1. Derivation of the Pendulum Equation

$$\frac{d^2\theta}{dt^2} + \frac{g}{L}\text{Sin}[\theta] = 0$$

The force acting on the pendulum can be broken into two components, one in the direction of the rod and the other in the direction of the pendulum's motion. If the rod does not stretch, the component directed along the rod plays no part in the pendulum's motion because it is counterbalanced by the force in the rod. The component in the tangential direction causes the pendulum to move. If we impose a coordinate system centered at the point of suspension of the pendulum and let $\mathbf{X}[\theta[t]]$ denote the vector giving the position of the pendulum bob, then Figure 46.1 illustrates the force of gravity on the pendulum and the resolution of this force into two components $\mathbf{F}_R$ and $\mathbf{F}_T$.

We will use the vector form of Newton's law, $\mathbf{F} = m\mathbf{A}$, to determine the equation of motion of the pendulum. We assume that the origin of our fixed inertial coordinate system is at the pivot point of the pendulum shown in the figure. The axes have their usual orientation.

HINT 46.2. *Radial and Tangential Position*
Express the position vector $\mathbf{X}[\theta[t]]$ of the mass in terms of the displacement angle θ, given that the length of the pendulum is L. Write $\mathbf{X}$ in terms of the unit vector

$$\mathbf{U}_R[\theta] = \begin{bmatrix} +\text{Sin}[\theta] \\ -\text{Cos}[\theta] \end{bmatrix}$$

If the vector $\mathbf{U}_R$ points in the direction of the rod, what are the directions of the x and y axes for the angle θ as shown in the figure?

Show that

$$\frac{d\mathbf{U}_R}{d\theta} = \mathbf{U}_T = \begin{bmatrix} \text{Cos}[\theta] \\ \text{Sin}[\theta] \end{bmatrix} \quad \text{and} \quad \frac{d^2\mathbf{U}_R}{d\theta^2} = -\mathbf{U}_R = \begin{bmatrix} -\text{Sin}[\theta] \\ +\text{Cos}[\theta] \end{bmatrix}$$

Prove that $\mathbf{U}_T$ is perpendicular to the rod. Prove that $\mathbf{U}_T$ points tangent to the motion of the rod. Sketch the pair $\mathbf{U}_R$ and $\mathbf{U}_T$ for $\theta = 0, +\pi/6, -\pi/4, +\pi/2, -4\pi/3, \pi$. Draw $\mathbf{U}_T$ with its tail at the tip of $\mathbf{U}_R$.

The next step in finding the equations of motion of the pendulum is to use some basic vector geometry. The gravity vector is simple in x-y coordinates, but we need to express it in radial and tangential components. You should be able to remember a simple entry in your geometric-algebraic vector lexicon of text Chapter 15 that lets you compute these components as perpendicular projections.

The magnitude of the gravitational force on the bob of mass m is mg and it acts down. The constant g is the universal acceleration due to gravity (which we discussed in Galileo's law in Chapter 10.) $g = 9.8$ in mks units. As a vector, the force due to gravity is

$$\mathbf{F}_G = mg \begin{bmatrix} +0 \\ -1 \end{bmatrix} = \begin{bmatrix} 0 \\ -mg \end{bmatrix}$$

HINT 46.3. *Radial and Tangential Gravity*
Decompose this vector $\mathbf{F}_G$ into the two vectors $\mathbf{F}_R$, the radial component of gravity, and $\mathbf{F}_T$, the tangential component of gravity. Express both vectors in terms of the vectors $\mathbf{U}_R[\theta]$ and $\mathbf{U}_T[\theta]$ defined above. Verify that your decomposition satisfies

$$\mathbf{F}_G = (?)\mathbf{U}_R + (??)\mathbf{U}_T = \mathbf{F}_R + \mathbf{F}_T$$

for the scalar quantities you compute.

The derivative of a general position vector with respect to t is the associated velocity vector. The second derivative of the position vector with respect to t is the acceleration vector.

HINT 46.4. *Radial and Tangential Acceleration*
Show that the velocity of the position $\mathbf{X}[\theta[t]]$ is given by

$$\frac{d\mathbf{X}}{dt} = \frac{d[L\mathbf{U}_R[\theta[t]]]}{dt} = L\frac{d\theta[t]}{dt}\mathbf{U}_T[\theta[t]]$$

You can verify this by writing $\mathbf{X}$ in sine-cosine components or by using Hint 46.2.

Calculate the second derivative the position vector $\mathbf{X}[\theta[t]]$ with respect to t and express the result in terms of the unit vectors $\mathbf{U}_R$ and $\mathbf{U}_T$. You will need to use the chain rule and product rule in symbolic form, since the angle θ depends on t and we do not have an explicit formula for $\theta[t]$. Write your answer in the form

$$\mathbf{A} = \frac{d^2\mathbf{X}}{dt^2} = (???)\mathbf{U}_R + (????)\mathbf{U}_T = \mathbf{A}_R + \mathbf{A}_T$$

We can now use $\mathbf{F} = m\mathbf{A}$ to determine the equation of motion of the pendulum. We first need to note that the rod from which the bob is suspended provides a force that counteracts the radial component of gravity and prevents any acceleration in the radial direction. Hence when we equate $\mathbf{F}$ and $m\mathbf{A}$, the radial component of the acceleration is exactly balanced by the force of tension in the rod and the two expressions cancel. Thus to determine the equation of motion of the pendulum we need only equate the expressions m times the tangential component of acceleration and the tangential component of gravity, $m\mathbf{A}_T = \mathbf{F}_T$.

HINT 46.5. *Combining Results*
Equate these two expressions and use the formulas above to derive the pendulum equation:

$$\frac{d^2\theta}{dt^2} + \frac{g}{L}\operatorname{Sin}[\theta] = 0.$$

Why doesn't the equation of motion for the pendulum depend on the mass? What does this mean physically?

46.2. Numerical Solutions of the Pendulum Equation

In order to solve second-order differential equations numerically, we must introduce a phase variable. If we let $\phi = \frac{d\theta}{dt}$, then the pendulum equation can be written as the system of differential equations:

$$\frac{d\theta}{dt} = \phi$$
$$\frac{d\phi}{dt} = -\frac{g}{L}\sin[\theta]$$

This system can then be solved by the computer program **AccDEsoln**. Also, this system is autonomous. Why? That means that we can use the **Flow2D** program to study the behavior of the pendulum.

HINT 46.6. *Equilibria*
What are the equilibrium points of the system above? Which equilibrium points are stable? Which are unstable? You should be able to determine this on purely physical grounds and then verify your conjectures with the comoputer. (What happens if you put the pendulum near the top?)

Figure 46.7 illustrates the phase plane trajectories of solutions to the system of differential equations for various initial conditions. The displacement angle θ is plotted on the horizontal axis, and the angular velocity ϕ is plotted on the vertical axis.

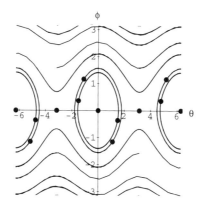

FIGURE 46.7: Phase Diagram for the Pendulum Equation

HINT 46.8. *An Invariant for the Pendulum*
Calculate an invariant for the pendulum motion by forming the ratio

$$\frac{\frac{d\phi}{dt}}{\frac{d\theta}{dt}} = -\frac{g\sin[\theta]}{L\phi}$$

canceling dt, separating variables, and integrating.

Prove that $E[\phi[t], \theta[t]] = \frac{L}{2}\phi^2[t] + g(1 - \cos[\theta[t]])$ is constant on solutions of the pendulum equations.

Make a contour plot of the function $E[\phi, \theta]$.

(HINT: See the main text, Section 24.4.)

EXPLORATION 46.9. *Mathematical and Physical Experiments*
Modify the **Flow2D** program to build a flow solution of the pendulum equations. We suggest that you use initial conditions ($L = 9.8$ meter, $g = 9.8$ m/sec^2).
$(\theta, \phi) = initxys = $ **Table[{x,0},{x,-2 Pi,2 Pi,.6}]**
and
$(\theta, \phi) = initxys = $ **Table[{0,x},{x,-2 Pi,2 Pi,.6}]**
As you can see from the phase plane there are two types of motion that the pendulum displays, the motion corresponding to the closed flow trajectories and that corresponding to the oscillating flow curves. Explain physically what type of motion is occurring when the flow trajectories are closed curves. Explain the type of motion occurring when the flow trajectories oscillate without closing on themselves. Use the **AccDEsoln** program to plot explicit time solutions associated with each kind of flow line.

We suggest that you also try some calculations with various lengths. A 9.8 m pendulum is several stories high and oscillates rather slowly. It would be difficult to build such a large pendulum and have it swing through a full rotation. How many seconds is a full oscillation (that is, what is the period)? How does the period depend on the length?

We suggest that you try some physical experiments as well, at least for small oscillations. With small variations in θ, a string will suffice for the rod. These experiments are easy to do and you should measure the period of oscillation for various lengths.

EXPLORATION 46.10. *The Period of the Pendulum*
Verify physically and by numerical experiments that the period depends on the length but not on the mass.

We would like to have a formula for the period, so we imagine the situation where we release the pendulum from rest at an angle $\alpha < \pi/2$. For the time period while θ is a decreasing function, say one-quarter oscillation, $0 \leq \theta \leq \alpha$, we could invert the function $\theta[t]$, $t = t[\theta]$. (See Project 21 and note that $1/\frac{d\theta}{dt} = \frac{dt}{d\theta}$, so $1/\phi = \frac{dt}{d\theta}$.) Now we proceed with some more clever tricks.

Compute the derivative

$$\frac{d(\phi^2)}{d\theta} = 2\phi \frac{d\phi}{d\theta} = 2\phi \frac{d\phi}{dt}\frac{dt}{d\theta} = 2\phi \frac{d\phi}{dt}\frac{1}{\phi}$$
$$= 2\frac{d\phi}{dt} = 2\frac{d^2\theta}{dt^2}$$

Use the pendulum equation and integrate with respect to θ

$$\frac{d(\phi^2)}{d\theta} = -2\frac{g}{L}L\ \text{Sin}[\theta]$$
$$(\phi)^2 = \frac{2g}{L}\int_\alpha^\theta (-\text{Sin}[\theta])\ d\theta \qquad \text{release from rest, } \phi[\alpha] = 0$$
$$\frac{d\theta}{dt} = \phi = \sqrt{\frac{2g}{L}}\sqrt{\text{Cos}[\theta] - \text{Cos}[\alpha]}$$

Finally, separate variables and integrate to the bottom of the swing, one quarter period,

$$dt = -\sqrt{\frac{L}{2g}} \frac{d\theta}{\sqrt{\text{Cos}[\theta] - \text{Cos}[\alpha]}}$$

$$T = 4 \int_\alpha^0 dt[\theta] = -4\sqrt{\frac{L}{2g}} \int_\alpha^0 \frac{d\theta}{\sqrt{\text{Cos}[\theta] - \text{Cos}[\alpha]}} d\theta$$

This integrand is discontinuous at $\theta = \alpha$ and may cause your computer trouble. We do some trig,

$$\text{Cos}[\theta] = \text{Cos}\left[\frac{\theta}{2} + \frac{\theta}{2}\right] = 1 - 2\left(\text{Sin}\left[\frac{\theta}{2}\right]\right)^2$$

$$\text{Cos}[\alpha] = 1 - 2\left(\text{Sin}\left[\frac{\alpha}{2}\right]\right)^2$$

Now

$$-\frac{1}{\sqrt{2}} \int_\alpha^0 \frac{1}{\sqrt{\text{Cos}[\theta] - \text{Cos}[\alpha]}} d\theta = -\int_\alpha^0 \frac{1}{\sqrt{\text{Sin}[\alpha/2]^2 - \text{Sin}[\theta/2]^2}} \frac{d\theta}{2}$$

$$= \int_0^{\pi/2} \frac{1}{\sqrt{1 - (\text{Sin}[\alpha/2])^2 \text{Sin}[\psi]}} d\psi$$

with the change of variables $\text{Sin}[\theta/2] = \text{Sin}[\alpha/2] \text{Sin}[\psi]$ and differentials $\text{Cos}[\theta/2] \, d\theta/2 = \text{Sin}[\alpha/2] \text{Cos}[\psi] \, d\psi$, with $\psi = 0$ when $\theta = 0$ and $\psi = \pi/2$ when $\text{Sin}[\theta/2] = \text{Sin}[\alpha/2] \text{Sin}[\psi]$, using $\text{Cos}[\psi] = \sqrt{1 - \text{Sin}[\psi]^2}$, we get finally,

$$T[\alpha] = 4\sqrt{\frac{L}{g}} \int_0^{\pi/2} \frac{1}{\sqrt{1 - (\text{Sin}[\alpha/2])^2 \text{Sin}[\psi]}} d\psi$$

$$= 8\sqrt{\frac{1}{1 - \text{Sin}[\alpha/2]^2}} \sqrt{\frac{L}{g}} \text{EllipticF}\left[\frac{\pi}{4}, -\frac{2\text{Sin}[\alpha/2]^2}{1 - \text{Sin}[\alpha/2]^2}\right]$$

This is an "elliptic integral of the first kind," or in *Mathematica* jargon, "EllipticF." The integral cannot be expressed in terms of elementary functions, but the computer can work with this expression perfectly well. See Figure 46.12 below.

46.3. Linear Approximation to the Pendulum Equation

If the displacement angle θ is small, then $\text{Sin}[\theta] \approx \theta$ and we can approximate the pendulum equation by the simpler differential equation:

$$\frac{d^2\theta}{dt^2} + \frac{g}{L}\theta = 0$$

This is a second-order linear constant-coefficient differential equation and can be solved explicitly for given initial conditions using the methods of text Chapter 23.

EXPLORATION 46.11. *The Linearized Pendulum's Period*
Show that the period of the linearized pendulum is a constant $2\pi\sqrt{\frac{L}{g}}$.

The true period of the pendulum differs from this amount more and more as we increase the initial release angle.

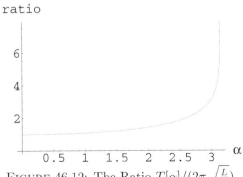

FIGURE 46.12: The Ratio $T[\alpha]/(2\pi\sqrt{\frac{L}{g}})$.

HINT 46.13. *Given initial values $\theta = \theta_0$ and $\frac{d\theta}{dt} = \phi_0$, solve the linearized pendulum equation symbolically. What are the solutions if the pendulum has no initial angular velocity, $\phi_0 = 0$. From the explicit solution, determine how long it takes for the linearized pendulum to swing back and forth if it is displaced an angle θ_0 and released. Does the period of the pendulum depend on the angle θ_0?*

As you demonstrated above the period of the linearized pendulum is a constant independent of the initial displacement (and mass). This should give you some idea of why pendula are such accurate timekeepers. If the pendulum is released at a certain angle to start and friction causes the amplitude of the swings to diminish, it still swings back and forth in the same amount of time. In a clock, friction is overcome by giving the pendulum a push when it begins to slow down. The push can come from weights descending, springs uncoiling, or a battery.

The comments about the invariance of period apply to the linearized model of the pendulum. That model is not accurate for large initial displacements.

HINT 46.14. *Comparison of Periods for Explicit Solutions*
Use the **AccDEsoln** program to solve the nonlinear pendulum equation for initial displacements running from 0 to $\frac{\pi}{2}$. Use $g = 9.8$ and $L = 9.8$ for the acceleration due to gravity and the length of the pendulum. The period of the linearized pendulum is a constant $2\pi\sqrt{\frac{L}{g}}$. How do the periods of the nonlinear pendulum compare to this constant? At what initial displacement does the linear approximation begin to break down? What happens if you vary the length?

Another way to compare the periods is in the **Flow2D** solutions.

HINT 46.15. *Flow of the Linearized Pendulum*
Use the computer program **Flow2D** to solve the equations

$$\frac{d\theta}{dt} = \phi$$
$$\frac{d\phi}{dt} = -\frac{g}{L}\theta$$

with $L = g = 9.8$ and initial conditions
(θ, ϕ) = **initxys** = **Table[{0,x},{x,-2 Pi,2 Pi,.6}]**

Compare the flow of the linearized model to the flow of the "real" (rigid rod frictionless) pendulum. Why do the dots of current state remain in line for the linear equation and not remain in line for the nonlinear flow?

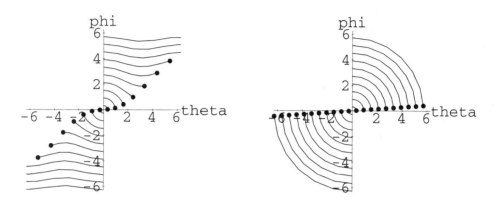

Nonlinear and Linear Pendula

The basis for our linear approximation to the equation of motion for the pendulum is the approximation $\sin[\theta] \approx \theta$.

HINT 46.16. *The Sine Approximation*
Recall the increment equation defining the derivative

$$f[x + \delta x] = f[x] + f'[x] \cdot \delta x + \varepsilon \cdot \delta x$$

where $\varepsilon \approx 0$ if $\delta x \approx 0$. Use this equation with $x = 0$, $\delta x = \theta$ and $f[x] = \sin[x]$ to give a first approximation to the meaning of "$\sin[\theta] \approx \theta$."

In the Mathematical Background there is a higher order increment equation, Taylor's second-order "small oh" formula,

$$f[x + \delta x] = f[x] + f'[x] \cdot \delta x + \frac{1}{2} f''[x] \cdot \delta x^2 + \varepsilon \cdot \delta x^2$$

Substitute into this equation with $x = 0$, $\delta x = \theta$, and $f[x] = \sin[x]$. Why does this give you a better approximation to the meaning of "$\sin[\theta] \approx \theta$"?

The Taylor formula can be used to show that

$$|f[x + \delta x] - f[x] - f'[x] \cdot \delta x| \leq \frac{1}{2} Max[|f''[\xi]| : x - \delta x \leq \xi \leq x + \delta x] \cdot \delta x^2$$

Use this to show that $|\sin[\theta] - \theta| \leq \theta^2/2$. If a pendulum swings no more than $10°$, how large is the error between $\sin[\theta]$ and θ? What is the relative error compared to the maximum angle? How does this compare with the relative error of 1 minute per week? One minute per month?

HINT 46.17. *Small Nonlinear Oscillations*
Use the computer program **Flow2D** to solve the equations

$$\frac{d\theta}{dt} = \phi$$

$$\frac{d\phi}{dt} = -\frac{g}{L} \text{Sin}\,[\theta]$$

with $L = g = 9.8$ and initial conditions
$(\theta, \phi) = \mathbf{\textit{initxys}} = \mathbf{\textit{Table}}[\{0,x\},\{x, "-10°", "10°", 0.02\}]$
How nearly do the dots remain in line for the nonlinear flow under only "10°" oscillations?

How many minutes per month might you expect for a pendulum clock that has oscillations varying from ten degrees fully wound to 8 degrees at the end of its regular winding period? Remember that the clock can be adjusted so that 9 degree oscillations would produce perfect time even though they do not have period $2\pi\sqrt{\frac{L}{g}}$.

46.4. Friction in the Pendulum (Optional)

How would you modify the pendulum equation to account for friction on the pivot of the pendulum? How does this affect the accuracy of a clock?

How would you modify the pendulum equation to account for air friction as the pendulum swings? How does this affect the accuracy of a clock?

46.5. The Spring Pendulum (Optional)

Now we consider a mass suspended from a spring instead of a rod. During the motion of the pendulum we will assume that the spring remains straight. The force of gravity still acts on the mass and can be resolved into tangential and radial components. There is also a force now due to extension of the spring. If the spring remains straight it will act in the radial direction, but its effect is dependent on whether the spring is stretched or compressed. Figure 46.18 illustrates the forces acting on the mass and their resolution into radial and tangential components.

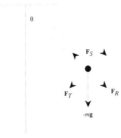

FIGURE 46.18: The Spring Pendulum

HINT 46.19. As with the simple pendulum, let $\mathbf{X}[\theta[t]]$ represent the position vector of the mass. The length of the pendulum now varies with time. Let $L(t)$ represent the length of the pendulum at time t. Express the vector $\mathbf{X}[\theta[t]]$ in terms of $L(t)$ and the unit vector $\mathbf{U}_R[\theta]$ defined above. The force from the spring obey's Hooke's law. If the unstretched length of the pendulum is L_0, then the magnitude of the spring force is $k(L(t) - L_0)$, where k is the spring constant. Express the spring force, $\mathbf{F}_S$, in terms of the unit vector $\mathbf{U}_R$ defined above. Resolve the force of gravity into radial and tangetial components, $\mathbf{F}_R$ and $\mathbf{F}_T$ as for the simple pendulum.

HINT 46.20. The second derivative of the position vector with respect to t is again the acceleration vector. Differentiate the expression for $\mathbf{X}[\theta[t]]$ twice with respect to t to obtain the acceleration vector. Express the acceleration vector in terms of the unit vectors $\mathbf{U}_R$ and $\mathbf{U}_T$. Be very careful. In this case both the length of the pendulum and the displacement angle are functions of t.

HINT 46.21. We again want to use Newton's law to determine the equation of motion of the pendulum. The total force acting on the mass is $\mathbf{F}_T + \mathbf{F}_R + \mathbf{F}_S$. In the exercise above, you have expressed this in terms of the unit vectors $\mathbf{U}_R$ and $\mathbf{U}_T$. You have also expressed the acceleration vector in terms of $\mathbf{U}_R$ and $\mathbf{U}_T$. Equate $\mathbf{F}$ and $m\mathbf{A}$ to obtain the equations of motion for the spring pendulum:

$$\frac{d^2 L}{dt^2} - L \left(\frac{d\theta}{dt} \right)^2 - g \cos[\theta] + \frac{k}{m}(L - L_0) = 0$$

$$L \frac{d^2\theta}{dt^2} + 2 \frac{dL}{dt} \frac{d\theta}{dt} + gL \sin[\theta] = 0$$

In order to solve the system of second-order differential equations numerically, we must again introduce phase variables to reduce it to a system of first-order differential equations. If we let $y_1 = L$, $y_2 = \frac{dL}{dt}$, $y_3 = \theta$, and $y_4 = \frac{d\theta}{dt}$, then the system becomes:

$$y_1' = y_2$$

$$y_2' = y_1 y_4^2 + g \cos[y_3] - \frac{k}{m}(y_1 - L_0)$$

$$y_3' = y_4$$

$$y_4' = -2y_2 y_4 / y_1 - g \sin[y_3] / y_1$$

HINT 46.22. Modify the **AccDEsoln** program to solve this system numerically. Use $g = 9.8$, $L_0 = 1.0$, $k = 9.0$, and $m = 1.0$ for the acceleration due to gravity, the unstretched length of the spring, the spring constant, and the mass of the bob. Start the pendulum by stretching the spring and giving it a small angular displacement. For example, take $y_1(0) = 2.5$ and $y_3(0) = 0.05$. Use 0 as initial conditions for y_2 and y_4. How does the amplitude of the swings change as time goes on? In particular, does the amplitude of the swings increase, decrease, or remain unchanged?

HINT 46.23. Associated with the spring pendulum are two "natural frequencies," $\omega_1 = \sqrt{\frac{g}{L_0}}$ and $\omega_2 = \sqrt{\frac{k}{m}}$. If the ratio of these frequencies is a rational number like 1, 2, or $\frac{1}{2}$, then the vibrations of the pendulum can exchange energies. Change the value of k so that $\frac{\omega_1}{\omega_2} = 1$ and use the same initial conditions as for the problem above. How does the amplitude of the swings change in this case? Change k so that $\frac{\omega_1}{\omega_2} = 2$ and perform the same experiment. Compare the amplitudes of the swings to the previous solutions. Finally, change k so that $\frac{\omega_1}{\omega_2} = \frac{5}{7}$ and perform the same experiments.

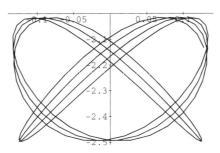

FIGURE 46.24: A Lissajous

HINT 46.25. *To get a better idea of how the vibrations change energy, we can plot the path that the bob takes as it moves on the end of the spring. The resulting graph will be what is sometimes called a lissajous. A lissajous is generated by plotting a superposition of two periodic motions. In this instance the two periodic motions are the vibration of the spring and the swinging of the pendulum. Figure 46.24 illustrates a sample path of the pendulum bob. Use the **Lissajous** program to generate the paths of the bob for the solutions you generated. Compare the sizes and styles of the different paths.*

PROJECT 47

Using Jupiter as a Slingshot

In order to provide a spacecraft with sufficient energy to escape the solar system, Jupiter can be used to "boost" its orbit. We want to determine the necessary relationship between the spacecraft's original trajectory and Jupiter's orbit, so that Jupiter will boost the spacecraft out of the solar system. This requires that they pass close enough for Jupiter to exert a definite effect, while not close enough for Jupiter to "catch" the spacecraft. To determine the relationship, we must find the general equations of motion.

We take the origin of the coordinate system at the sun. In order to simplify matters, the solar system is represented in two dimensions, with the coordinate axes in the plane of the orbits of the planets. Figure 47.1 is a diagram of the system. In the diagram, the vector **X** points from the sun to the spacecraft and the vector $\mathbf{X}_J$ points from the sun to Jupiter (the inner circle represents the earth's orbit, while the outer circle represent Jupiter's orbit).

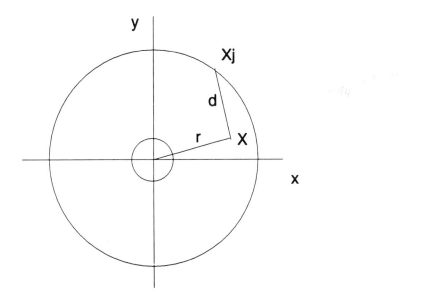

FIGURE 47.1: Solar System in Two Dimensions

47.1. Setting up the Problem: Scaling and Units

It is convenient (even necessary for accuracy) to use units of measurement that simplify the calculations. The general unit of astronomical length is the AMU (astronomical unit), where 1 AMU = the approximate distance from the earth to the sun. The distance from Jupiter to the sun is 5.2 AMU. Mass is measured in a system based on the sun as the unit; the sun is defined as having mass 1. In comparison, then, the mass of Jupiter is 0.001, and the mass of the spacecraft is tiny. As an additional convenience, time is defined in a manner so that the gravitational constant $G = 1$ in Newton's law of gravity,

$$f = G \frac{M\,m}{r^2}$$

One time unit is $\frac{365}{2\pi}$ earth days.

VARIABLES

t = time measured from launch of the spacecraft

$\mathbf{X} = \begin{bmatrix} x \\ y \end{bmatrix}$ = the vector position of the spacecraft

x = x − coordinate of the spacecraft

y = y − coordinate of the spacecraft

$\mathbf{X}_J = \begin{bmatrix} x_J \\ y_J \end{bmatrix}$ = the vector position of Jupiter

x_J = x − coordinate of Jupiter

y_J = y − coordinate of Jupiter

d = distance from Jupiter to the spacecraft

r = distance from the sun to the spacecraft

HINT 47.2. *The important quantities r and d are secondary variables. Use your geometry-algebra lexicon to show that*

$$r = \sqrt{x^2 + y^2}$$
$$d = \sqrt{(x - x_J)^2 + (y - y_J)^2}$$

Hints: What is r in terms of the vector $\mathbf{X}$? What is r in terms of the components of $\mathbf{X}$? Sketch the vector $\mathbf{X} - \mathbf{X}_J$. What are the components of this vector? What is its length in components?

Kepler's third law states that the square of the period of a planet is proportional to the cube of its mean distance to the sun, specifically,

$$P_J^2 = \frac{4\pi^2}{GM} r^3$$

The period and frequency of Jupiter's orbit about the sun are given in these units by

$$P_J = 2\pi\,(5.2)^{3/2}$$
$$n_J = 2\pi/P_J = 1/(5.2)^{3/2}$$

where $m_J = 0.001$ is the mass of Jupiter. We shall assume that the spacecraft is so small that it does not disturb the circular orbit of Jupiter. It is a simple matter to obtain the following parametric representation for the orbit of Jupiter:

$$x_J = 5.2 \operatorname{Cos}[n_j(t - t_0)]$$
$$y_J = 5.2 \operatorname{Sin}[n_j(t - t_0)]$$

where t_0 depends on the initial position of Jupiter.

HINT 47.3. *Review the parametric equations of a circle from Section 16.2 of the main text and verify that these equations describe Jupiter's circular orbit.*
When $t = 0$, where is Jupiter?
How long does it take for Jupiter to complete one revolution about the sun?

47.2. Newton's Law of Gravity

Newton's law of gravity says that an object of mass m and a second object of mass M attract each other with a force of magnitude

$$G \frac{m \cdot M}{r^2}$$

where r is the distance between the objects. The force acts along a line connecting the centers of the objects.

HINT 47.4. *Let m_s denote the mass of the spacecraft and let m_J denote the mass of Jupiter. Recall that $\mathbf{X}$ is the vector position of the spacecraft (as a function of time) and $\mathbf{X}_J$ is the vector position of Jupiter.*

(1) *Show that the vector $-\frac{1}{r}\mathbf{X}$ is a unit length vector pointing from the spacecraft toward the sun.*
(2) *Show that Newton's gravitational law says the force on the spacecraft due to the sun is*

$$\mathbf{F}_{sun} = -G \frac{m_s \, m_{sun}}{r^3} \mathbf{X}$$

(3) *Which vector points from the spacecraft to Jupiter?*
(4) *Show that the vector $\frac{1}{d}(\mathbf{X} - \mathbf{X}_J)$ is a unit length vector. In what direction does it point?*
(5) *Show that Newton's gravitational law says that the force on the spacecraft due to Jupiter is*

$$\mathbf{F}_{Jup} = -G \frac{m_s \, m_J}{d^3} (\mathbf{X} - \mathbf{X}_J)$$

47.3. Newton's $F = ma$ Law

Newton's "F = m a" law can now be applied to find the equations of motion of the spacecraft:

$$m_s \mathbf{a} = \mathbf{F} \quad \text{the total force on the craft}$$

$$m_s \frac{d^2 \mathbf{X}}{dt^2} = \mathbf{F}_{sun} + \mathbf{F}_{Jup}$$

$$m_s \frac{d^2 \mathbf{X}}{dt^2} = -\frac{m_s m_J}{r^3} \mathbf{X} - \frac{m_s m_J}{d^3} (\mathbf{X} - \mathbf{X}_J)$$

$$\frac{d^2 \mathbf{X}}{dt^2} = -\frac{1}{r^3} \mathbf{X} - \frac{m_J}{d^3} (\mathbf{X} - \mathbf{X}_J)$$

HINT 47.5. *Use vector components to re-write the equations of motion of our spacecraft as*

$$\frac{d^2 x}{dt^2} = -\frac{x}{r^3} - m_J \frac{(x - x_J)}{d^3}$$

$$\frac{d^2 y}{dt^2} = -\frac{y}{r^3} - m_J \frac{(y - y_J)}{d^3}$$

In order to use this system of equations, we need to interpret our original equations as a corresponding set of four first-order differential equations, which can then be evaluated and plotted as functions of time using the **JupiterHelp** program. You learned the "phase variable trick" in Chapter 23 of the core text, and so the derivation is left to you.

HINT 47.6. *Rewrite the system as a system of four first-order differential equations.*

47.4. Numerical Flights out of the Solar System

Use the **JupiterHelp** program to find initial conditions that will "sling" your spacecraft out of the solar system.

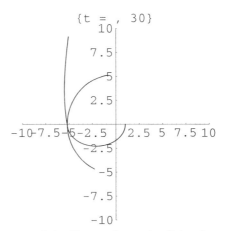

FIGURE 47.7: Escape from the Solar System

Chapter 15

Chemical Reactions

PROJECT 48

Stability of a Tank Reaction

This project models a chemical reaction where cold chemicals pour into a tank and react. It is based on papers of a number of authors beginning with R. Aris and N. R. Amundson in 1958 in the chemical engineering journals. We use data of Berger and Perlmutter from 1964. Scaling is a problem with most of the real data.

The reaction in the tank heats the chemicals and they pour out the other side of the tank at new concentrations. Some of the input chemical flows out the other side, but at a reduced concentration. That reduction has produced the substance we want to manufacture.

The reaction produces heat, but the cold chemical flowing in can absorb some heat. If it absorbs too much, the reaction stops. If it absorbs too little, things overheat. We want to know conditions for a stable equilibrium, in other words, conditions where the reaction continues at a stable temperature.

48.1. Mass Balance

VARIABLES

t = time measured from the start of the reaction in hours

c = the concentration of the input chemical in the tank at time t in pound-moles

T = the temperature in degrees Fahrenheit

We assume that the tank has volume V and is full from the start. This means that if chemical flows in at volume rate q (cubic feet/hr), then it also flows out at volume rate q. If the concentration of the chemical coming in is c_{in}, then

THE AMOUNT OF CHEMICAL COMING IN PER HOUR IS

$q \cdot c_{in}$ - cubic feet per hour × chemical per cubic foot = chemical per hour

The concentration in the tank is c and it is flowing out at the rate q, so

THE AMOUNT OF CHEMICAL GOING OUT PER HOUR IS

$q \cdot c$ - cubic feet per hour × chemical per cubic foot = chemical per hour

One more thing happens to our input chemical: it reacts to form something else, thereby disappearing. The reaction rate depends on the concentration and the temperature.

48.2. Arrhenius' Law

Arrhenius' Law says that amount of chemical disappearing per unit volume per hour is given by
$$A e^{-B/T} c$$
for constants A and B. The amount of chemical disappearing is thus $V\, A e^{-B/T} c$.

HINT 48.1. *Show that the rate of change of the amount of chemical in the tank is*
$$V \frac{dc}{dt} = q\, c_{in} - q\, c - V\, A e^{-B/T} c$$
or that the rate of change of the concentration of the input chemical in the tank is given by
$$\frac{dc}{dt} = p(c_{in} - c) - c\, A e^{-B/T}$$
What is the parameter p?

We have used a number of parameters in the description and we need a few more.

PARAMETERS

q = flow rate in and out of the tank – units?
c_{in} = flow rate in and out of the tank – units?
T_{in} = the incoming temperature in °F
V = the volume of the tank in cubic feet
A = Arrhenius' constant
B = Arrhenius' other constant
p = q/V
H = the rate of heat generation of the reaction in BTU per pound mole
s_C = the specific heat of the chemical in BTU/ft^3 °F

48.3. Heat Balance

We need to account for the total amount of heat. The specific heat of the chemical is the parameter s_C. This means that a unit volume of chemical at temperature T holds $s_C T$ BTUs. This tells us two amounts

THE AMOUNT OF HEAT COMING IN IS
$$q\, s_C\, T_{in}$$

THE AMOUNT OF HEAT GOING OUT IS
$$q\, s_C\, T[t]$$

Finally, we need to know the amount of heat generated. Recall that Arrhenius' Law says that the amount of reaction per unit volume is $A e^{-B/T} c$, so there is a total amount of reaction $V A e^{-B/T} c$. The parameter H measures the rate of production of heat per unit of reaction, so

THE AMOUNT OF HEAT GENERATED IS

$$H V A e^{-B/T} c$$

HINT 48.2. *Show that the time rate of change of the amount of heat in the tank is*

$$V s_C \frac{dT}{dt} = q s_C T_{in} - q s_C T + H V A e^{-B/T} c$$

and find the derivative of temperature with respect to time $\frac{dT}{dt} =??$

In short, we have found nonlinear differential equations of the form:

$$\frac{dc}{dt} = f[c, r]$$
$$\frac{dr}{dt} = g[c, r]$$

where $x = c$ and $r = T/T_{in}$.

HINT 48.3. *Show that*

$$f[c, r] = p(c_{in} - c) - c\, e^{a-b/r}$$
$$g[c, r] = p(1 - r) + c\, h\, e^{a-b/r}$$

where $r = T/T_{in}$ *and the computed constants are*

$$p = q/V$$
$$a = Log[A]$$
$$b = \frac{B}{T_{in}}$$
$$h = \frac{H}{s_C T_{in}}$$

Measured parameters from the literature are:

parameter	Berger	Aris	Luyben	test case
q	$= 200$	$= 100$	$= 2000$	$= 100$
c_{in}	$= 0.270$	$= 1$	$= .50$	$= 1$
T_{in}	$= 530$	$= 350$	$= 530$	$= 300$
V	$= 100$	$= 100$	$= 2400$	$= 100$
A	$= 10^8$	$= e^{25}$	$= 7.08\ 10^{10}$	$= e^{25}$
B	$= 10^4$	$= 10^4$	$= 15000$	$= 9000$
H	$= 10^4$	$= 200$	$= 600$	$= 200$
s_C	$= 50$	$= 1$	$= .75$	$= 1$

We will use the test case values for a sample computation. (The program **TankReaction** on our website is based on the text program **Flow2D**.) The flow of the differential equations in this case looks as in Figure 48.4.

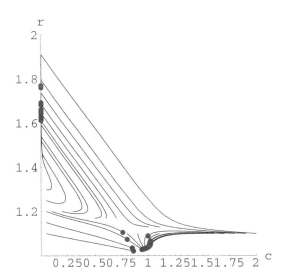

FIGURE 48.4: A Flow Solution of the Tank Reaction Equations

Notice that there is an attracting equilibrium at a high temperature about 1.7 times the input temperature and another near the input temperature where $c \approx 1$. The high-temperature equilibrium is using up most of the chemical and the low-temperature one is the case where the reaction has died out. We want to find conditions that guarantee that the equilibrium that is using up input chemical is stable – neither explosively growing nor dying out. We use Theorem 24.6 from the main text.

48.4. Stability of Equilibria

Our first job is to solve the equations for an equilibrium with the test case values of the parameters.

$$f[c,r] = p\,(c_{in} - c) - c\;e^{a-b/r} = 0$$
$$g[c,r] = p\,(1-r) + c\,h\;e^{a-b/r} = 0$$

Your computer probably cannot solve these nonalgebraic equations exactly, so we begin with a plot of the implicit curves $f[c,r] = 0$ and $g[c,r] = 0$.

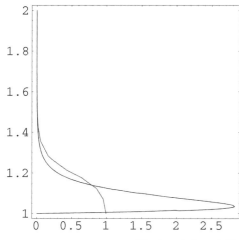
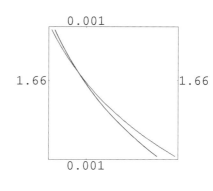

Horizontal and Vertical Vector Fields

We used a numerical root-finding method to locate the three crossing points as follows: $(0.992191, 1.00521)$, $(0.790976, 1.13935)$, and $(0.000917083, 1.66606)$.

Next, we apply Theorem 24.6 of the main text. We take

$$\begin{bmatrix} \frac{\partial f}{\partial x}[c_E, r_E] & \frac{\partial f}{\partial y}[c_E, r_E] \\ \frac{\partial g}{\partial x}[c_E, r_E] & \frac{\partial f}{\partial y}[c_E, r_E] \end{bmatrix} = \begin{bmatrix} f_x & f_y \\ g_x & g_y \end{bmatrix}$$

The characteristic equation is

$$\text{Determinant}\left[\begin{bmatrix} f_x - x & f_y \\ g_x & g_y - x \end{bmatrix}\right] = x^2 + (-f_x - g_y)\,x + (f_x\,g_y - f_y\,g_x) = 0$$

$$= x^2 + \beta\,x + \gamma = 0$$

HINT 48.5. (1) *Show that*

$$\beta = 2p + (1 - \frac{b\,h\,c_E}{r_E^2})\,e^{a - b/r_E}$$

and

$$\gamma = p\left(p + (1 - \frac{b\,h\,c_E}{r_E^2})\,e^{a - b/r_E}\right)$$

(2) *Show that the real part of the two solutions to the characteristic equation*

$$x^2 + \beta\,x + \gamma = 0$$

are both negative if and only if both β and γ are positive.

(3) *Use this fact and Theorem 24.6 to show that the equilibrium at $(0.790976, 1.13935)$ is unstable, but the equilibria at $(0.992191, 1.00521)$ and $(0.000917083, 1.66606)$ are stable for the test case values of the parameters.*

The program TankReaction on our website, based on **Flow2D**, may help you make a broader investigation of the tank dynamics.

EXPLORATION 48.6. *Investigate the reactions corresponding to the parameter values from the literature.*

48.5. Forced Cooling (Optional)

EXPLORATION 48.7. *Suppose we could devise a cooling mechanism that sensed the temperature of the reacting chemicals and supplied cooling, say*

$$V\, K[T] = \text{the rate of heat removal supplied by your cooler}$$

Derive the new set of differential equations

$$f[c, r] = p\,(c_{in} - c) - c\, e^{a-b/r}$$
$$g[c, r] = p\,(1 - r) + c\, h\, e^{a-b/r} + k[r]$$

where $k[r] = K[r\, T_{in}]/(T_{in}\, s_C)$ and recompute the stability conditions of the equations in terms of the new function $k[r]$.

Experiment on some unstable reactions to see how to run your cooler to stabilize them.

PROJECT 49

Beer, Coke, & Vitamin C

Project 35 discusses linear drug models that accurately model many drugs. Beer, cocaine, and vitamin C are exceptions. Ethanol and cocaine are metabolized by an enzyme reaction and absorption of vitamin C is nonlinear. All three are well modeled by a limiting case of the general enzyme equations known as the Michaelis-Menten equation.

49.1. Enzyme-mediated Reactions

"Enzymes" are chemicals that bind to a "substrate" and allow an irreversible reaction to take place creating a "product." The reaction frees the enzyme to do its job again but does not use it up. The substrate will not react, or only react very slowly, unless it is bound by the enzyme. Schematically, our enzyme reaction looks as follows:

$$Z + S \underset{k}{\overset{h}{\rightleftarrows}} C \overset{r}{\rightarrow} P + Z$$

where S represents substrate, Z represents enzyme, C represents the substrate-enzyme complex, and P represents the product.

49.2. Molar Concentration and Reaction Rates

We want to write equations that describe how these reactions proceed. The amounts of the various substances change in time, so time derivatives of a measure of each will give us such a description. In this case a good choice of units to measure the substances will greatly simplify the description. A natural choice might be to measure mass, say in grams, or the mass concentration of each substance, since a typical reaction takes place in liquid. This is not the best choice. Units of molar concentration make our reaction balances easiest to describe with derivatives.

A mole of a compound is Avagadro's number, 6.022×10^{23}, of molecules of the substance. This is the atomic weight of the compound in grams, so, for example, a mole of carbon 12 weighs 12 grams. We will use units of "molars" or moles per liter in our examples, specifically millimolars. The reason this choice of units is best is that one molecule of enzyme combines with one molecule of substrate to form one molecule of complex.

VARIABLES

t = time measured from the start of the reaction in seconds
z = the concentration of enzyme Z at time t in millimolars
s = the concentration of substrate S at time t in millimolars
c = the concentration of complex at time t in millimolars
p = the concentration of complex at time t in millimolars

The choice of molar concentrations means that the rates on the left side of the schematic reaction satisfy
$$\frac{dc}{dt} = -\frac{ds}{dt}$$
and on the right side
$$-\frac{dc}{dt} = \frac{dp}{dt}$$
Enzyme is "lost" or bound on the left and "gained" or released on the right.

PARAMETERS

h = the reaction rate of disassociation of complex
k = the reaction rate of enzyme and substrate forming complex
r = the reaction rate of complex forming product and free enzyme

Each of the rate constants is proportional to the product of the concentration of the reactants, so

$k\,s\,z$ = the rate of the forward reaction of enzyme and substrate
$h\,s\,z$ = the rate of the backward reaction of enzyme and substrate
$r\,c$ = the rate of the forward reaction of complex to product

As a differential equation the enzyme-substrate forms and unforms by
$$\frac{ds}{dt} = -k\,s\,z + h\,c$$
The reaction of complex to product is given by
$$\frac{dp}{dt} = r\,c$$
Three things change the concentration of complex. It can disassociate back into substrate and enzyme, it can react into product and enzyme, or substrate and enzyme can combine to form it:
$$\frac{dc}{dt} = k\,s\,z - h\,c - r\,c$$
Since enzyme is either free and part of z or bound and part of c,
$$-\frac{dz}{dt} = \frac{dc}{dt} = k\,s\,z - h\,c - r\,c$$

HINT 49.1. *Some New Constants, a and b, and the General Dynamics*

(1) *Prove that the sum $a = s + c + p$ is constant. Hint: The derivative of a constant is zero.*
(2) *Prove that $b = z + c$ is constant.*
(3) *Use these relationship to prove that the four equations for $\frac{ds}{dt}$, $\frac{dz}{dt}$, $\frac{dc}{dt}$, and $\frac{dp}{dt}$ reduce to just the equations*

$$\frac{dp}{dt} = r\, c$$
$$\frac{dc}{dt} = k\,(b - c)(a - p - c) - (h + r)\, c$$

Explain how you would find $z[t]$ and $s[t]$ if you were given a solution to these two equations.

(4) *How are a and b determined at the beginning of a reaction? Write your answer as an augmented initial value problem,*

$$z[0] = z_0, \quad s[0] = s_0, \quad c[0] = 0, \quad p[0] = 0$$
$$a = ???, \quad b = ???$$
$$\frac{dp}{dt} = r\, c$$
$$\frac{dc}{dt} = k\,(b - c)(a - p - c) - (h + r)\, c$$

49.3. The Briggs-Haldane Dynamics Approximation

In some reactions, after a short period, the NET rate of formation of complex is small compared with the flow of substrate to complex and complex to product, so $\frac{dc}{dt} \approx 0$.

HINT 49.2. *The Briggs-Haldane Initial Value Problem*

(1) *If $\frac{dc}{dt} \approx 0$, show that*

$$c \approx \frac{b\, s}{H + s}$$

where $H = (h + r)/k$. (Hint: Use the constant b and the equation for the change in c.)

(2) *In this case, show that*

$$\frac{dp}{dt} = r\, \frac{b\, s}{H + s}$$

since $c \approx$ constant.

(3) *Now use the constant $a = s + c + p$ to show that $\frac{ds}{dt} \approx -\frac{dp}{dt}$. This gives the Briggs-Haldane approximation:*

$$s[0] = s_0$$
$$\frac{ds}{dt} = -r\, \frac{b\, s}{H + s}$$

(4) *How can we find $p[t]$ and $z[t]$ from the solution of the Briggs-Haldane IVP?*

EXPLORATION 49.3. *Briggs-Haldane Numerical Experiments*
Compare the dynamics of the solution of the full enzyme dynamics in Hint 49.1 with the Briggs-Haldane initial value problem in the case where $h =?$, $k =?$, $r =?$ and in the case $h =??$, $k =??$, $r =??$

49.4. The Michaelis-Menten Dynamics Approximation

Usually, we use a great deal of substrate and only a little enzyme to perform a reaction. The Michaelis-Menten approximation arises when the amount of substrate is large compared to the amount of enzyme and the product reaction is slow compared with the disassociation, $r \ll h$. In this case, $s \gg z$, c would rather go back to s and z, so c and p are small. Thus, the constant $a = s + c + p \approx s$. We always have $b = z + c$, constant, so $z = b - c$.

If k and h are large compared with r, the complex used to produce product will be quickly replaced and c will stay nearly constant. While the net concentration of c is approximately in equilibrium,

$$\frac{dc}{dt} = k(b-c)(a-p-c) - (h+r)c$$
$$0 \approx k(b-c)(a) - (h+r)c$$
$$0 \approx k(b-c)(s) - hc$$

HINT 49.4. *Michaelis-Menten Initial Value Problem*
Solve the equation
$$0 = -k(b-c)s + hc$$
for c and show that
$$c = \frac{bs}{K+s}$$
for the constant $K = \frac{h}{k}$.

(1) Substitute this into the rate formation equation for product and show
$$\frac{dp}{dt} = r\frac{bs}{K+s}$$

(2) Recall that substrate turns into product to obtain the Michaelis-Menten initial value problem,
$$s[0] = s_0$$
$$\frac{ds}{dt} = -r\frac{bs}{K+s}$$

(3) How do you determine p[t] from the solution of the Michaelis-Menten initial value problem?

A LIMITING CASE OF MICHAELIS-MENTEN DYNAMICS
Since the constant $b = z + c$, or $c = b - z$, and the rate of reaction forming product is $\frac{dp}{dt} = rc$, the maximal rate of production is

$$\text{Maximal rate of production} = \rho_{Max} = rb$$

Also, in general,
$$\frac{dp}{dt} = r\frac{bs}{K+s} \quad \text{so} \quad \frac{dp}{dt} = \frac{\rho_{Max}\, s}{K+s}$$
In real kinetic experiments, scientists plot $1/\frac{dp}{dt}$ versus $1/s$ to measure the constants.

HINT 49.5. *Show that $1/\frac{dp}{dt}$ varies linearly with $1/s$,*
$$\frac{1}{\frac{dp}{dt}} = \frac{K}{\rho_{Max}} \frac{1}{s} + \frac{1}{\rho_{Max}}$$
and explain how you could measure K from an experimental graph of $1/\frac{dp}{dt}$ versus $1/s$.

Ask your local neighborhood biochemist for some real examples where the Briggs-Haldane and Michaelis-Menten approximations apply. Specifically, could $k \ll h$ when $h \gg r$? Will c remain nearly constant for most of a reaction even when r, h, and k are comparable, but z is small?

HINT 49.6. *Comparison of Briggs-Haldane and Michaelis-Menten Dynamics*
How do the Briggs-Haldane and Michaelis-Menten approximations compare? For example, under the Michaelis-Menten assumptions, how do the constants $H = \frac{h+r}{k}$ and $K = \frac{h}{k}$ compare? Under the Michaelis-Menten assumptions, can the Briggs-Haldane assumptions also hold? What about the converse?

Mathematically, you can explore the differences:

EXPLORATION 49.7. *Michaelis-Menten Numerical Experiments*
Compare the dynamics of the solution of the full enzyme dynamics in Hint 49.1 and the Briggs-Haldane approximation ?? with the Michaelis-Menten initial value problem in the case where $h = ?$, $k = ?$, $r = ?$ and in the case $h = ??$, $k = ??$, $r = ??$

49.5. Blood Ethanol

Blood alcohol (ethanol) at high concentrations like 100 millimolars causes vomiting and inability to walk. Alcohol is metabolized by an enzyme called "alcohol dehydrogenase." The constant K for this enzyme is about 3 millimolars.

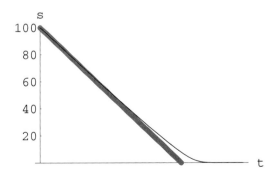

FIGURE 49.8: Sobering up for 4 Hours

HINT 49.9. *Hockey Sticks*

(1) If $\frac{K}{s} \approx 0$ for most of the decay of a substance, show that

$$\frac{s}{K+s} \approx 1$$

so that the differential equation is initially approximately

$$\frac{ds}{dt} \approx ?$$

with solution $s[t] = s_0 - r\,b\,t$.

(2) Compare this solution with the Michaelis-Menten dynamics with $r = 50$, $K = 3$, and $b = 0.01$.

(3) Why do the pharmacokinetics books call these curves "hockey stick curves"?

EXPLORATION 49.10. *Look up the "linear" sobering rate for blood ethanol and adjust the constants above appropriately.*

49.6. Blood CO_2

Carbon dioxide (CO_2) is closely related to blood pH (acidity). There is an approximately constant concentration of CO_2 in the blood of about 1.35 millimolars. One mechanism that keeps this level constant is a reaction with an enzyme caled "carbonic anhydrase." The constant K for this enzyme is about 12 millimolars. The average concentration of the enzyme is about 0.001 and the rate constant is on the order of 10^6 per second, one of the largest enzyme rate constants.

HINT 49.11. *Exponential Approximation*

(1) If $s \ll K$, we can approximate the Michaelis-Menten equation by

$$\frac{ds}{dt} = -r\frac{b}{K}s$$

Why?

(2) Show that the solution to the approximate equation is $s[t] = s_0\, e^{-r\,b\,t/K}$.

(3) Compare the exponential solution with the solution to the Michaelis-Menten initial value problem when $K = 12$ and $r\,b = 1000$.

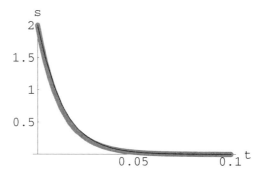

FIGURE 49.12: CO_2 Concentration in the Blood

EXPLORATION 49.13. *A Visit to the Chemistry Library*

Find more exact values of the constants for the carbonic anhydrase reaction.

Find an intermediate speed reaction where Michaelis-Menten dynamics lie between hockey sticks and exponentials.

Chapter 16

More Mathematical Projects

PROJECT 50

Rearrangement of Conditionally Convergent Series

Dirichlet's theorem 13.4 from the Mathematical Background book (on the CD) gives us the convergent expression

$$\frac{x}{2} = \text{Sin}[x] - \frac{1}{2}\text{Sin}[2\,x] + \frac{1}{3}\text{Sin}[3\,x] - \frac{1}{4}\text{Sin}[4\,x] + \cdots + \frac{(-1)^{k+1}}{k}\text{Sin}[k\,x] + \cdots$$

but notice that the simple absolute value estimate of Weierstrass' majorization from Section 25.2 of the main text leads to no conclusion when we estimate

$$\left|\frac{(-1)^{k+1}}{k}\text{Sin}[k\,x]\right| \leq \frac{1}{k}$$

since the harmonic series diverges,

$$1 + \frac{1}{2} + \frac{1}{3} + \cdots + \frac{1}{n} \to \infty$$

Dirichlet's theorem for Fourier series is very powerful, because it gives us results where great care is needed for direct convergence proofs. (The proof of Dirichlet's theorem uses integration by parts somewhat like the proof of Taylor's formula.)

We emphasized absolute uniform convergence in the main text, because that kind of convergence allows us to treat series the most like very long sums. This section shows you an example of the strange nature of series that do not converge absolutely. Actually, we could expand on the specific idea and show that for any number r or $r = \pm\infty$, we can rearrange the terms of a conditionally convergent series so that it sums to r. By rearranging terms, we can get *any* sum we choose! (The proof of this is not very hard, but it is more abstract than the specific example we give.)

We showed in Example 27.1 of the main text that

$$1 - \frac{1}{2} + \frac{1}{3} + \frac{1}{4} + \cdots + \frac{(-1)^{n+1}}{n} \to \text{Log}[2]$$

converges and that it even has an error no more than $\frac{1}{n+1}$, since the approximating sums alternate up and down with decreasing size oscillations. However, we want you to observe that rearranging these same terms results in a different limit, specifically,

313

Prove that:

$$1 - \frac{1}{2} + \frac{1}{3} - \frac{1}{4} + \cdots + \frac{(-1)^{n+1}}{n} + \cdots \neq 1 + \frac{1}{3} - \frac{1}{2} + \frac{1}{5} + \frac{1}{7} - \frac{1}{4} + \cdots$$

HINT 50.1. *Show that the rearranged alternating harmonic series above may be written symbolically as*

$$\text{Sum}\left[\left(\frac{1}{4k-3} + \frac{1}{4k-1} - \frac{1}{2k}\right), \{k, ?, \infty\}\right] =$$

$$1 - \frac{1}{3} - \frac{1}{2} + \frac{1}{5} + \frac{1}{7} - \frac{1}{4} + \cdots + \left(\frac{1}{4k-3} + \frac{1}{4k-1} - \frac{1}{2k}\right) + \cdots$$

HINT 50.2. *Prove that the grouped terms $(1+\frac{1}{3}) - \frac{1}{2} + (\frac{1}{5}+\frac{1}{7}) - \frac{1}{4} + \cdots$ produce decreasing oscillations, that is,*

$$\left(\frac{1}{4k+3} + \frac{1}{4k+1}\right) < \frac{1}{2k} < \left(\frac{1}{4k-3} + \frac{1}{4k-1}\right)$$

HINT 50.3. *Also prove that each of the terms grouped as in the first expression is positive,*

$$\left(\frac{1}{4k-3} + \frac{1}{4k-1} - \frac{1}{2k}\right) > 0$$

and use this to prove that the rearranged series converges and

$$1 + \frac{1}{3} = \frac{4}{3} > 1 + \frac{1}{3} - \frac{1}{2} + \frac{1}{5} + \frac{1}{7} - \frac{1}{4} + \cdots$$

$$1 + \frac{1}{3} - \frac{1}{2} = \frac{5}{6} < 1 + \frac{1}{3} - \frac{1}{2} + \frac{1}{5} + \frac{1}{7} - \frac{1}{4} + \cdots$$

$$1 + \frac{1}{3} - \frac{1}{2} + \frac{1}{5} + \frac{1}{7} = \frac{247}{210} > 1 + \frac{1}{3} - \frac{1}{2} + \frac{1}{5} + \frac{1}{7} - \frac{1}{4} + \cdots$$

$$1 + \frac{1}{3} - \frac{1}{2} + \frac{1}{5} + \frac{1}{7} - \frac{1}{4} = \frac{389}{420} < 1 + \frac{1}{3} - \frac{1}{2} + \frac{1}{5} + \frac{1}{7} - \frac{1}{4} + \cdots$$

HINT 50.4. *Clearly, the terms of the alternating series*

$$1 - \frac{1}{2} + \frac{1}{3} - \frac{1}{4} + \cdots$$

sum with decreasing oscillations, so prove that

$$1 > 1 - \frac{1}{2} + \frac{1}{3} - \frac{1}{4} + \cdots$$

$$\frac{1}{2} < 1 - \frac{1}{2} + \frac{1}{3} - \frac{1}{4} + \cdots$$

$$\frac{5}{6} > 1 - \frac{1}{2} + \frac{1}{3} - \frac{1}{4} + \cdots$$

$$\frac{7}{12} < 1 - \frac{1}{2} + \frac{1}{3} - \frac{1}{4} + \cdots$$

$$\frac{47}{60} > 1 - \frac{1}{2} + \frac{1}{3} - \frac{1}{4} + \cdots$$

HINT 50.5. *Combine your results to show*

$$1 - \frac{1}{2} + \frac{1}{3} - \frac{1}{4} + \cdots < \frac{5}{6} < 1 + \frac{1}{3} - \frac{1}{2} + \frac{1}{5} + \frac{1}{7} - \frac{1}{4} + \cdots$$

$$1 - \frac{1}{2} + \frac{1}{3} - \frac{1}{4} + \cdots < \frac{47}{60} < \frac{5}{6} < \frac{389}{420} < 1 + \frac{1}{3} - \frac{1}{2} + \frac{1}{5} + \frac{1}{7} - \frac{1}{4} + \cdots$$

$$\vdots$$

(Note: $\frac{389}{420} - \frac{47}{60} = \frac{1}{7} \approx 0.14$.)

You are welcome to use the symbolic summation of the computer. For example, to compare these series with *Mathematica*, you could type the program:

T := Sum[(1/(4 k - 3) + 1/(4 k - 1) - 1/(2 k) ,{k , 1, n}] ;

S := Sum[1/(2 j - 1) - 1/(2 j) , { j , 1, n}] ;

Do[

Print["T = " , T , "S = " , S , "n = " , n, "Diff approx =", N[T - S]]

,{ n , 1, 10 }]   *< Enter >*

PROJECT 51

Computation of Fourier Series

Fourier series and general "orthogonal function expansions" are important in the study of heat flow and wave propagation as well as in pure mathematics. The reason that these series are important is that sines and cosines satisfy the "heat equation" or "wave equation" or "Laplace's equation" for certain geometries of the domain. A general solution of these partial differential equations can sometimes be approximated by a series of the simple solutions by using superposition. (Fourier series provide many interesting examples of delicately converging series because of Dirichlet's theorem 13.4 of the Mathematical Background.)

The method of computing Fourier series is quite different from the methods of computing power series. The Fourier sine-cosine series associated with $f[x]$ for $-\pi < x \leq \pi$ is:

$$f[x] \sim a_0 + \text{Sum}[a_k \, \text{Cos}[kx] + b_k \, \text{Sin}[kx], \{k, 1, \infty\}]$$

where

$$a_0 = \frac{1}{2\pi} \int_{-\pi}^{\pi} f[x] \, dx = \text{Average of } f[x]$$

and for $k = 1, 2, 3, \cdots$

$$a_k = \frac{1}{\pi} \int_{-\pi}^{\pi} f[x] \cdot \text{Cos}[kx] \, dx \quad \text{and} \quad b_k = \frac{1}{\pi} \int_{-\pi}^{\pi} f[x] \cdot \text{Sin}[kx] \, dx$$

Dirichlet's theorem says that if $f[x]$ and $f'[x]$ are 2π-periodic and continuous except for a finite number of jumps or kinks and if the value $f[x_j]$ is the midpoint of the jump if there is one at x_j, then the Fourier series converges to the function at each point. It may not converge uniformly, in fact, the approximating *graphs may not converge to the graph of the function*, as in Gibbs' goalposts below. If the periodic function $f[x]$ has no jumps (but may have a finite number of kinks, or jumps in $f'[x]$), then the series converges uniformly to $f[x]$.

Convergence of Fourier series is typically weaker than the convergence of power series, as we shall see in the examples, but the weak convergence is still quite useful. Actually, the most important kind of convergence for Fourier series is "mean square convergence,"

$$\int_{-\pi}^{\pi} [f[x] - S_n[x]]^2 \, dx \to 0$$

where $S_n[x]$ is the sum of n terms. This is only a kind of average convergence, since the integral is still small if the difference is big only on a small set of $x's$. We won't go into mean square convergence except to mention that it sometimes corresponds to the difference in energy between the "wave" $f[x]$ and its approximation $S_n[x]$. Mean square convergence has important connections to "Hilbert spaces."

Convergence of Fourier series at "almost every" point was a notorious problem in mathematics, with many famous mathematicians making errors about the convergence. Fourier's work was in the early 1800s and not until 1966 did L. Carleson prove that the Fourier series of any continuous function $f[x]$ converges to the function at almost every point. (Dirichlet's theorem uses continuity of $f'[x]$. Mean square convergence is much easier to work with and was well understood much earlier.)

Three basic examples of Fourier sine-cosine series are animated in the computer program **FourierSeries**. These are the following.

$$f[x] = |x| \quad \text{for} \quad -\pi < x \leq \pi, \quad \text{repeated } 2\pi \text{ periodically}$$

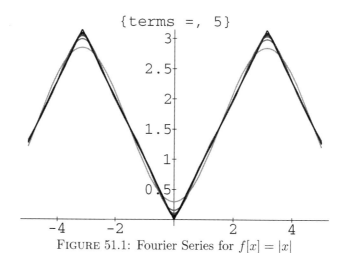

FIGURE 51.1: Fourier Series for $f[x] = |x|$

The average value of $f[x]$ is clearly $\pi/2$ and can be computed as the integral

$$a_0 = \frac{1}{2\pi} \int_{-\pi}^{\pi} f[x] \, dx = \frac{1}{2\pi} \int_{-\pi}^{\pi} |x| \, dx$$

$$= 2 \frac{1}{2\pi} \int_{0}^{\pi} x \, dx$$

$$= \frac{1}{\pi} \frac{1}{2} x^2 \Big|_0^\pi = \frac{\pi^2}{2\pi}$$

$$= \frac{\pi}{2}$$

Notice the step in the computation of the integral where we get rid of the absolute value. We must do this in order to apply the Fundamental Theorem of Integral Calculus. Absolute value does not have an antiderivative. We do the same thing in the computation of the other coefficients.

$$a_{2k} = \frac{1}{\pi} \int_{-\pi}^{\pi} |x| \cos[2k\,x]\, dx$$

$$= \frac{2}{\pi} \int_0^{\pi} x \cos[2k\,x]\, dx$$

$$= \frac{2}{\pi} \left[x \frac{1}{2k} \sin[2kx] \Big|_0^{\pi} - \int_0^{\pi} \frac{1}{2k} \sin[2k\,x]\, dx \right]$$

$$= \frac{2}{\pi} \left[2k\pi \sin[2k\pi] - 0 - \frac{1}{(2k)^2} (\cos[2kx] \Big|_0^{\pi}) \right] = 0$$

using integration by parts with

$$u = x \qquad\qquad dv = \cos[2kx]\, dx$$
$$du = dx \qquad\qquad v = \frac{1}{2k} \sin[2kx]$$

HINT 51.2. *Show that the a_k terms of the Fourier series for $f[x] = |x|$ with odd k are*

$$a_{2k+1} = -\frac{4}{\pi} \cdot \frac{1}{(2k+1)^2}$$

and all $b_k = 0$.

Without the absolute value the integrals of the Fourier coefficients can be computed directly, without breaking them into pieces.

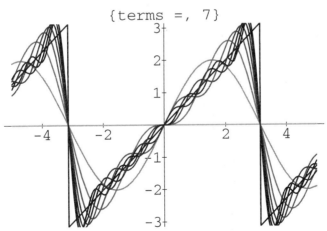

FIGURE 51.3: Fourier Series for $f[x] = x$

HINT 51.4. *Calculate the coefficients of the Fourier sine-cosine series for $f[x] = x$, for $-\pi < x \leq \pi$, extended to be 2π periodic. Notice that the average $a_0 = 0$, by inspection of the graph or by computation of an integral. Moreover, $x \cos[2kx]$ is an odd function, that* is, $-x\operatorname{Cos}[2k\cdot(-x)] = -(x\operatorname{Cos}[2kx])$, so the up areas and down areas of the integral cancel, $a_k = 0$. You can compute this from the formula. Finally, show that

$$b_k = 2\frac{(-1)^{k+1}}{k}$$

The coefficients for the Fourier series of

$$f[x] = \operatorname{Sign}[x] = \begin{cases} +1, & \text{if } x > 0 \\ 0, & \text{if } x = 0 \\ -1, & \text{if } x < 0 \end{cases}$$

must be computed by breaking the integrals into pieces where the Fundamental Theorem applies, for example,

$$a_0 = \frac{1}{2\pi}\int_{-\pi}^{\pi} f[x]\, dx = \frac{1}{2\pi}\left[\int_{-\pi}^{0} -1\, dx + \int_{0}^{\pi} +1\, dx\right] = 0$$

HINT 51.5. *Show that*

$$a_k = \frac{1}{\pi}\int_{-\pi}^{\pi} \operatorname{Sign}[x]\,\operatorname{Cos}[kx]\, dx = 0$$

and

$$b_{2k} = 0$$

because each piece of the integral $\int_0^\pi \operatorname{Sin}[2kx]\, dx = 0$, being the integral over whole periods of the sine function.

Finally, show

$$b_{2k+1} = \frac{2}{\pi}\int_0^\pi \operatorname{Sin}[(2k+1)x]\, dx$$

$$= \frac{2}{\pi}\cdot\frac{1}{2k+1}\int_0^{(2k+1)\pi} \operatorname{Sin}[u]\, du$$

$$= \frac{4}{\pi}\cdot\frac{1}{2k+1}$$

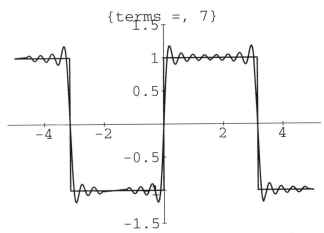

FIGURE 51.6: Gibbs Goalposts $f[x] = \text{Sign}[x]$

The convergence of the Fourier series for $\text{Sign}[x]$ holds at every fixed point, but the convergence is not uniform. In fact, the graphs of the approximations do not converge to a "square wave," but rather to "goalposts." Each approximating term has an overshoot before the jump in $\text{Sign}[x]$ and these move to a straight line segment longer than the distance between ± 1. You can see this for yourself in the animation of the computer program **FourierSeries**.

PROJECT 52

The Big Bite of the Subtraction Bug

When computers do arithmetic with fixed length decimal (or binary) approximations, numbers of vastly different sizes cannot be accurately subtracted. For example, one million and one, $1,000,001 = 0.1000001 \times 10^7$, represented to only six decimals is 0.100000×10^7. The difference between one million and one and one million is one, $1,000,001 - 1,000,000 = 1$, but if this is done with a six decimal machine, the difference is zero, $(0.100000 - 0.100000) \times 10^7 = 0$. In this example, the baby is thrown out with the bath water, and there is 100% loss of accuracy. This kind of loss of a significant portion of the small difference between two relative large numbers plagues numerical computation.

The one general kind of series whose error term is the size of the next term is an alternating series of decreasing size terms,

$$a_0 - a_1 + a_2 - a_3 + \cdots$$

where $|a_n| > |a_{n+1}|$. However, these series are not reliable to use in floating point computations, because of the subtraction bug. The computer can do symbolic computations and simplify symbolically before evaluating numerically, so there are various ways for you to try computations that involve close differences of large numbers.

HINT 52.1. *Experiments with $e^x = 1 + x + \cdots x^n/n! + \cdots$*

Here is one way to sum the series for the natural exponential:

T = 1;
S = 1;
x = -30.0;
a = Exp[x];
Do[
T = T x/k;
S = S + T;
Print["S =",S," k = ",k,"Sum = ",a,"T = ", T]
,{k,1,100}];

If the arithmetic is done with perfect accuracy, after 30 terms, the terms decrease and the error should be no more than the next term. At $k = 100$ you will see that the predicted accuracy is quite high, but you will also see that the estimate is wrong.

Compare this sum with $1/S$ when you run the program with $x = +30.0$. Since the sum with $x = +30$ is approximately e^{30} and since $1/3^{30} = e^{-30}$, these should be approximately equal. Are they actually?

Compare your summation with the computer's built-in algorithm, for example,

1/N[Exp[30],10]

N[Exp[-30],10]

Mathematica can do this sum with its built-in sum command in two ways,

x = 30;

N[Sum[x∧n/n! , {n,0,100}]]

or

x = 30.0;

Sum[x∧n/n! , {n,0,100}]

Also try to sum symbolically first with

Sym = Together[Sum[y∧n/n!,{n,0,100}]];

y = -30;

Print[Sym , N[Sym]]

The numerical sum methods will give different answers that are not accurate because of the subtraction bug, but the specifics depend on your particular machine!

Watch out for the subtraction bug, it could sneak up and bite your computer.

Chapter 17

Additional Project References

You may want to develop your own project if you have a science, engineering, or math professor who can help get you started.

1001.1. Lowering the Water Table

Recently a student asked me about a hydrology problem where drilling a well posed a threat to a nearby wetland. A simplified version of this problem would make a good project. We checked some references and I could see how to simplify the geometry for the aquifer so we could see some basic results. There wasn't time to finish that work before this went to e-press.

In math there are many books with titles like "Analysis of Applied Mathematical Models" to give you ideas, for example, *Mathematics for Dynamic Modeling* by Edward Beltrami, Academic Press, 1987.

"UMAP Modules" should also be available in your math library. These are similar to our projects – designed to show how undergraduate math is applied in a host of areas. The specific topics used in the modules are specified at the beginning of each module.

The book *Applications of Calculus*, Philip Straffin, editor, MAA Notes Number 29, Mathematical Association of America, 1993, has more projects. The robot arm looks like a good one.

The book *Student Research Projects in Calculus* by Cohen, Gaughan, Knoebel, Kurtz, and Pengelley, Mathematical Association of America, 1991, has still more.

Finally, I hope to expand and maintain a website for student projects in calculus.

1001.2. The Light Speed Lighthouse

For example, the lighthouse problem of Example CD 7.11 in the main text predicts that a beam of light will sweep down a straight shoreline faster than the speed of light. Of course, this isn't so, but the part of the shore where it matters is far out. The speed of the beam along the beach (in miles per second) is given by

$$\frac{dD}{dT} = \frac{\frac{\pi}{10} \operatorname{Sec}[\frac{\pi}{10}t]}{1 + \frac{\pi}{10} \operatorname{Sec}[\frac{\pi}{10}t] \operatorname{Tan}[\frac{\pi}{10}t]/c}$$

$$= \frac{\pi}{10}\left(1+D^2\right) \frac{1}{1 + \frac{\pi D \sqrt{1+D^2}}{10\,c}}$$

$$= c\, \frac{1}{\frac{1}{\sqrt{1+1/D^2}} + \frac{10\,c}{\pi(1+D^2)}}$$

But since $c = 18600$ miles per second, when D is a modest distance, the textbook answer is close, since $\frac{\pi D \sqrt{1+D^2}}{10\,c}$ is small and the second formula gives the approximate speed $\frac{\pi}{10}\left(1+D^2\right)$. When D tends to infinity, however, the limiting speed is c, as we see from the last formula. The computations are easiest to illustrate with the computer and we will put a program on the website as soon as possible.

1001.3. Horizontal, Vertical, and Slant Asymptotes

Another traditional topic that is pretty dry on paper and much more fun with a computer is the study of horizontal, vertical, and slant asymptotes to graphs. Again, look for a program on **Asymptotes**. Stay tuned.